SpringerBriefs in Electrical and Computer Engineering

More information about this series at http://www.springer.com/series/10059

Yuan Wu · Li Ping Qian · Jianwei Huang
Xuemin (Sherman) Shen

Radio Resource Management for Mobile Traffic Offloading in Heterogeneous Cellular Networks

 Springer

Yuan Wu
College of Information Engineering
Zhejiang University of Technology
Hangzhou, Zhejiang
China

Li Ping Qian
College of Information Engineering
Zhejiang University of Technology
Hangzhou, Zhejiang
China

Jianwei Huang
Department of Information Engineering
The Chinese University of Hong Kong
Shatin
Hong Kong

Xuemin (Sherman) Shen
Department of Electrical and Computer
  Engineering
University of Waterloo
Waterloo, ON
Canada

ISSN 2191-8112   ISSN 2191-8120   (electronic)
SpringerBriefs in Electrical and Computer Engineering
ISBN 978-3-319-51036-1   ISBN 978-3-319-51037-8   (eBook)
DOI 10.1007/978-3-319-51037-8

Library of Congress Control Number: 2016960558

Printed on acid-free paper

This Springer imprint is published by Springer Nature
The registered company is Springer International Publishing AG
The registered company address is: Gewerbestrasse 11, 6330 Cham, Switzerland

Preface

The past decade has witnessed an unprecedented growth of smart mobile devices and mobile Internet services, which leads to a tremendous increase of mobile traffic in cellular access networks. It becomes a critical challenge for the mobile network operators to efficiently accommodate such a heavy traffic demand. Conventional schemes such as directly upgrading network infrastructures and upgrading access technologies are undesirable due to the need of huge capital investments. By exploiting the multi-tier architecture of cellular networks, on the other hand, it becomes possible to actively offload mobile users' traffic from macrocells to heterogenous small cells, and hence accommodate the heavy traffic in a cost effective fashion. However, achieving the promising benefits of traffic offloading requires proper radio resource allocations, and thus in this brief, we aim at providing concise discussions regarding radio resource management schemes for traffic offloading in heterogenous cellular networks.

In Chap. 1, we start by providing an overview of the heterogenous cellular networks and traffic offloading. Then, we illustrate different traffic offloading paradigms in heterogeneous cellular networks. To illustrate the deigns of optimal radio resource allocations for traffic offloading in heterogeneous cellular networks, we provide two concrete design examples and present the corresponding promising benefits in Chaps. 2 and 3. Specifically, in Chap. 2, we study the paradigm of mobile users' uplink traffic offloading through small cells, and investigate the joint traffic scheduling and power allocation problem. The problem aims at minimizing mobile users' overall mobile data cost by properly offloading to small cells while avoiding severe co-channel interference. In Chap. 3, we study the paradigm of downlink traffic offloading via mobile users' device-to-device (D2D) cooperations for content distribution, and investigate the joint D2D-assisted users' cooperations and content transmission control problem with the objective of minimizing the overall radio resource usage. Finally, in Chap. 4, we draw conclusions and discuss about future research directions.

This brief illustrates the designs of optimal radio resource allocations for traffic offloading in heterogeneous cellular networks and exhibits the corresponding promising benefits. This brief can be used as a textbook or reference book for postgraduate students working on advanced topics in wireless communications and networking.

We would like to thank Mr. Yanfei He and Mr. Jiachao Chen from Zhejiang University of Technology for their research contributions to the presented Springer Brief. We also would like to thank all members of BBCR group at the University of Waterloo for their valuable discussions and insightful suggestions. Special thanks are due to the editors at Springer Science+Business Media, Melissa Fearon, Jennifer Malat, and Susan Lagerstrom-Fife for their help throughout the publication preparation process.

This work is supported in part by the National Natural Science Foundation of China (61572440, 61303235, 61379122), and in part by the Zhejiang Provincial Natural Science Foundation of China (LR17F010002, LR16F010003), and in part by the Young Talent Cultivation Project of Zhejiang Association for Science and Technology (2016YCGC011), and in part by the General Research Funds (Project Number CUHK 14202814 and 14219016) established under the University Grant Committee of the Hong Kong Special Administrative Region, China, and in part by the Natural Sciences and Engineering Research Council, Canada.

Hangzhou, China   Yuan Wu
Hangzhou, China   Li Ping Qian
Shatin, Hong Kong   Jianwei Huang
Waterloo, Canada   Xuemin (Sherman) Shen

Contents

1   **Traffic Offloading in Heterogeneous Cellular Networks**.............. 1
    1.1   Heterogeneous Small-Cell Cellular Networks.................. 1
    1.2   Challenging Issues in Heterogeneous Small-Cell Networks....... 2
    1.3   Traffic Offloading in Heterogeneous Small-Cell Networks........ 4
        1.3.1   Small-Cell-Based Traffic Offloading.................. 6
        1.3.2   D2D-Assisted Traffic Offloading..................... 8
    1.4   Open Issues in Traffic Offloading in Heterogeneous
        Small-Cell Networks 9
    1.5   Aim of the Brief 12
    References 12

2   **Resource Allocation for Small-Cell-Based Traffic Offloading** 17
    2.1   Related Studies 18
    2.2   System Model and Problem Formulation.................... 19
        2.2.1   System Model 19
        2.2.2   Problem Formulation 21
    2.3   Equivalent Problem Transformations 22
    2.4   Efficient Algorithm for Optimal Offloading Solution 25
        2.4.1   Layered Structure.................................. 25
        2.4.2   Algorithm for Optimal Offloading Solution 27
    2.5   Numerical Results 30
    2.6   Extension for Bandwidth Allocation 36
    2.7   Summary 39
    References 39

3   **Resource Allocation for D2D-Assisted Traffic Offloading** 43
    3.1   Related Studies 44
    3.2   System Model 45
        3.2.1   MUs' Cooperative Offloading Through D2D-Links 45
        3.2.2   Models of Resource Consumption 47

3.3 Problem Formulation and Decomposable Structure 49
    3.3.1 Problem Formulation 49
    3.3.2 Decomposable Structure 50
3.4 Optimal Offloading Solution and Proposed Algorithm 52
    3.4.1 Optimal Offloading Duration: Solution of Subproblem 52
    3.4.2 Optimal Content Transmission: Solution
          of Top-Problem 53
    3.4.3 Algorithm for Optimal Offloading Solution 59
3.5 Numerical Results 60
3.6 Extension for Distributing Multiple Pieces of Contents 67
    3.6.1 System Model and Problem Formulation 67
    3.6.2 Illustrative Approach for Optimal Solution 70
3.7 Summary 71
References 71

4 Conclusions and Future Directions 75
4.1 Conclusions 75
4.2 Future Research Directions 76
References 77

Acronyms

3GPP	The Third Generation Partnership Project
AAA	Authentication, Authorization, Accounting
ABSs	Almost Blank Sub-frames
AP	Access Point
BS	Base Station
CAGR	Compound Annual Growth Rate
D2D	Device-to-Device
DC	Dual Connectivity
eICIC	Enhanced Inter-cell Interference Coordination
HCN	Heterogeneous Cellular Networks
LTB	Listen Before Talk
LTE	Long Term Evolution
LTE-A	Long Term Evolution Advanced
MU	Mobile User
OFDM	Orthogonal Frequency Division Multiplexing
OFDMA	Orthogonal Frequency Division Multiple Access
QoS	Quality of Service
RAN	Radio Access Networks
RB	Resource Block
SINR	Signal to Interference plus Noise Ratio
SON	Self Organizing Networks
SSL	Secure Socket Layer
TDMA	Time Division Multiple Access
UE	User Equipment

Chapter 1
Traffic Offloading in Heterogeneous Cellular Networks

Media-hungry mobile devices and mobile traffic have been experiencing an exponential growth in the past decade. As reported in Cisco Visual Networking Index (VNI) [1], the global mobile traffic grew from 2.1 exabytes/month at the end of 2014 to 3.7 exabytes/month at the end of 2015, corresponding to 74% growth in 2015, and the growth rate is expected to continue at a compound annual growth rate at 53% until 2020. The huge traffic demand has overloaded cellular radio access networks (RANs), which in comparison experience a much slower capacity increase. It becomes a critical challenge for the cellular operators to accommodate the heavy traffic demand in a timely and cost-efficient manner. Directly upgrading RANs may be undesirable from operators' perspective for the following two concerns: (i) upgrading RANs requires a huge capital investment, which may not be easily recovered even by accommodating the traffic demand, and (ii) acquiring more licensed spectrum bands for the upgraded RANs is difficult and expensive due to the regulation policy. Fortunately, nowadays cellular networks are structured in a multitier architecture, namely, a large number of heterogeneous small cells (such as picocells, femtocells, and WiFi systems) have densely underlaid conventional macrocells [2]. Hence, traffic offloading through small cells provides an effective and cost-efficient way to accommodate mobile users' traffic.

In this chapter, we first overview heterogeneous cellular networks (HCNs) and several challenging issues in HCNs. We next illustrate different paradigms of traffic offloading in HCNs. We then discuss about several open issues in traffic offloading and illustrate the contributions of this brief.

1.1 Heterogeneous Small-Cell Cellular Networks

Cellular systems have been paving the way towards the architecture comprised of multitier heterogeneous small cells, ready for the approaching 5G cellular systems [3]. In HCNs, in addition to conventional high-powered macrocells providing large

© The Author(s) 2017
Y. Wu et al., *Radio Resource Management for Mobile Traffic Offloading in Heterogeneous Cellular Networks*, SpringerBriefs in Electrical and Computer Engineering, DOI 10.1007/978-3-319-51037-8_1

coverage, a large number of low-powered small cells (such as femtocells, picocells, and WiFi access points (APs)) have been densely deployed, with the objective of bringing radio access networks (RANs) closer to mobile users (MUs). Thanks to the close proximity to MUs, small cells can ensure high-quality transmission links, and thus bring benefits such as improving coverage, enhancing throughput, and reducing radio resource consumption. In addition, the large number of available small cells provide a greater freedom for users' network associations and traffic delivery, hence improving quality of services (QoS) and radio resource utilization. There exists several types of small cells in HCNs, with different coverage, power capacity, and spectrum occupancy [2]:

- *Picocells*: Picocells are deployed by cellular operators, either regularly or irregularly, at traffic hot-spots to provide a medium-level coverage area ($\leq$300 m), with a transmit-power in the range of 250 mW–2 W. Picocells operate on licensed bands and use the X2 interface as backhaul [4].
- *Femtocells*: Femtocells are deployed by individuals mainly to provide indoor coverage ($\leq$50 m), with a transmit-power in the range of 10–100 mW [5]. Femtocells usually operate on licensed bands and use the broadband Internet connection as backhaul.
- *WiFi*: WiFi systems are small cells operating on unlicensed ISM bands (2.4 and 5 GHz bands). WiFi systems are deployed either by operators or individuals to provide a small coverage area, with a transmit-power in the range of 100–200 mw (e.g., IEEE 802.11b/g/n) [6, 7]. WiFi systems relay on the broadband Internet connection as backhaul.

In addition to the above different types of small cells, HCNs also include other paradigms such as relays and distributed antenna systems. Relay nodes are often deployed by cellular operators close to the edges of macrocells to extend the cells' coverage and capacity. Distributed antenna systems (DAS) refer to a network of spatially separated antenna nodes which are connected to a common transport medium (such as optimal fiber cable) to provide high capacity wireless coverage within an area [8].

1.2 Challenging Issues in Heterogeneous Small-Cell Networks

To reap the potential benefits of small cells, we need to design proper radio management mechanisms that involve several key challenging issues [2]. We provide a brief discussion about these issues in the following.

Intercell Interference Management: Intercell interference management is one of the most critical issues in HCNs. Due to the scarcity of licensed spectrum resource, cellular operators tend to reuse frequency when deploying small cells over licensed bands (such as picocells and femtocells) [9], which leads to severe intercell interference if not managed properly. Such intercell interferences, including both cross-tier interference between macrocells and small-cells and intra-tier interference among different small cells, have significantly limited the capacity expansion gain of small-cells deployment. For instance, severe macro-to-femto uplink interference will occur when an MU close to a femtocell AP transmits to the macrocell BS. Therefore, it is necessary to deploy intercell interference management schemes (such as the enhanced intercell interference coordination (eICIC) techniques in 3GPP standardization activities), to maximize the capacity expansion gain of densely deploying small cells. Several categories of eICIC techniques have been proposed, namely, the time-domain techniques such as the almost blank subframes used by femtocells, the frequency-domain techniques such as the strict (or soft) fractional frequency reuse, and the power control techniques [2]. Moreover, since users' cell-associations influence the intercell interference, joint consideration of cell association and eICIC techniques will further help us address the intercell interference issue.

Restricted Access-Mode Control: Since small cells such as femtocells and WiFi APs are allowed to be individually configured, the access mode control is an important issue. There exist two different access modes for those small cells, namely, (i) the open-access with which small cells allow arbitrary nearby MUs to access, and (ii) the closed-access with which small cells only allow authorized MUs to access and completely exclude unauthorized ones. Access-mode control of small cells will influence the network performance, such as throughput, availability of connection, and coverage. For instance, despite benefiting authorized MUs (e.g., in terms of experiencing less-congested backhaul links), closed-access mode adversely influences the unauthorized users' experiences due to unavailability of service. The optimal choice of access mode depends on several factors, such as the adopted orthogonal-based/nonorthogonal-based multiple access schemes and the MUs' density and distribution [10].

Cell Association and Handover Management: Cell association is one of the most crucial issues that influence the MUs' experienced quality of service (QoS). In an overlapping coverage area served by several different small cells, myopically associating an MU with the cell offering the largest signal-to-interference-plus-noise ratio (SINR) might not lead to an overall optimized network performance. Proper cell association schemes need to take into account several factors, such as link quality, users' QoS requirements, and cells' load levels and backhaul capacities [11].

Meanwhile, handover management, including the handover between macrocells and small cells and the handover among different small cells, becomes more important with dense small cells deployment, since moving MUs encounter cell edges more frequently. There already exist some applications supporting the automatic cell association and handover in today's cellular systems [12]. Nevertheless, to optimize the network-wise performance, we need to take into account several issues, e.g., the computational and signaling overheads, and the interference mitigation in handover process in dense small cells. For instance, studies in [13, 14] proposed efficient handover schemes that aim at reducing the number of handovers and mitigating the cross-tier interference in small cells.

Exploitation of Unlicensed Spectrums: As the number of small cells keep increasing, the scarcity of licensed spectrum resource has become a critical bottleneck for capacity expansion with more small cells. To address the issue and avoid the huge expanse in obtaining new licensed spectrums, recent 3GPP standardization activities and industrial proposals have been exploiting unlicensed spectrums for the operations of small cells. The vast existence of unlicensed spectrums (such as ISM bands in 2.4 and 5 GHz bands) are expected to boost the capacity of small cells significantly. However, exploiting unlicensed bands necessitates a premise that the incumbent users who already operate in unlicensed bands do not suffer from significant harmful inferences. In other words, a harmonic coexistence is required between the existing systems (such as Zigbee and Bluetooth) using unlicensed spectrums and the small cells which opportunistically exploit unlicensed spectrums [15]. To achieve this goal, protocols based on the rationale of listen-before-talk (LTB) have been proposed in 3GPP LTE-A regarding the use of unlicensed bands.

Network Controlled Device-to-Device Communications: In HCNs, network controlled device-to-device (D2D) communication is an emerging paradigm providing an underlay to cellular systems [16, 17]. D2D communications enable MUs of close proximity to communicate with each other directly, instead of through the BS. Thus, D2D communications can not only help reduce uplink/downlink resource consumptions, but also save the transmit-power consumptions due to the short distance between MUs. Integrating D2D communications into cellular systems requires new resource management schemes. Since D2D communications are managed by cellular BSs, the network operators need to consider both proper radio resource block (RB) allocations [18, 19] and proximity discovery schemes [20]. In addition, it is crucial to properly select between the D2D-mode and cellular-mode for MUs in order to optimize their transmission performance as well as the network-wise performance [21, 22]. Interference mitigation is also an important issue for D2D communications, since cellular networks usually reuse RBs for D2D communications for saving spectrum usage. The authors of [23, 24] took into account the interference due to RB-reuse and aimed at optimizing total throughput in D2D communications. Similarly, the authors of [25–27] aimed at optimizing the energy efficiency of D2D communications while accounting for the RB-reuse interference. Most studies considered inband (i.e., using the licensed bands of cellular system) RB-sharing approach to establish D2D-links, since the inband approach can yield better spectrum efficiency [28]. In comparison, outband (overlay) approach uses dedicated licensed bands for establishing D2D-links, which relieves interference between D2D-links and cellular links but lowers the spectrum efficiency.

1.3 Traffic Offloading in Heterogeneous Small-Cell Networks

The large number of small cells in HCNs provide high-quality links for flexibly offloading MUs' traffic from macrocells, yielding a cost-effective solution to accommodate the rapidly growing MUs' traffic. In particular, nowadays most smart wireless devices (such as smartphones and tablets) are equipped with multiple radio interfaces, namely, one long-range 3G/4G interface and at least one short-range interface such as WiFi and Bluetooth, which facilitate traffic offloading through small cells. Traffic offloading through small cells can bring multifold benefits to both cellular operators and MUs.

- **Benefits for Cellular Networks**: Thanks to the close-proximity to MUs, small cells can provide better link quality (with higher throughputs and lower congestions) than macrocells. Thus, traffic offloading to small cells not only can relieve traffic pressure in macrocells, but also can bring benefits such as enhancing throughput and coverage as well as reducing radio resource consumption. In particular, since some small cells (such as femtocells and WiFi systems) use broadband Internet as backhauls, traffic offloading through those small cells can reduce traffic burden at backhauls of cellular RANs.
- **Benefits for MUs**: Traffic offloading can benefit the MUs in terms of reducing mobile data cost and improving QoS. Specifically, small cells such as WiFi access points usually charge MUs at much lower prices and provide higher data rates than macrocells, which motivate the MUs to offload data to small cells. An MU can flexibly choose different types of small cells to satisfy different types of QoS.

The benefits of data offloading have lead to the proposals of several different paradigms, among which the two most important and prevalent paradigms are: (i) traffic offloading through small cells, which is also referred as the small-cell-based traffic offloading in the remainder of the brief, and (ii) traffic offloading based the MUs' proximity-cooperations via D2D-communications, which is also referred as the D2D-assisted traffic offloading. Figure 1.1 illustrates these two main paradigms. We illustrate the small-cell-based traffic offloading and the D2D-assisted traffic offloading in the next two subsections, respectively.

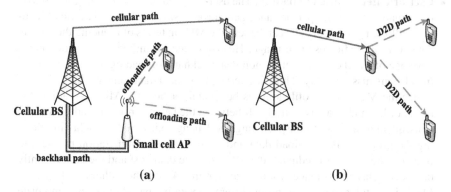

Fig. 1.1 Illustration of two categories of traffic offloading paradigm in heterogeneous small-cell networks. **a** small-cell-based traffic offloading; **b** D2D-assisted traffic offloading

1.3.1   Small-Cell-Based Traffic Offloading

Traffic offloading through small cells is one of the most prevalent paradigms for offloading MUs' traffic in HCNs. Studies about this paradigm also consists of two directions: (i) the network-oriented traffic offloading through small cells, and (ii) the user-oriented traffic offloading through small cells.

- **Network-oriented Traffic Offloading**: The network-oriented traffic offloading focuses on studying the overall network performance achieved by small-cell-based traffic offloading. Due to limited radio resources and potential intercell interferences, there usually exists a strong coupling among users when offloading happens. For example, severe intercell interference occurs when nearby small cells offload traffic on the same frequency bands. In addition, severe interference and collision occur when many MUs aggressively offload data to the same small cell. Hence, it is important to understand the network-wise performance gain from small-cell-based traffic offloading. In [29], Ho et al. considered the intercell interferences among different small cells accommodating the offloaded traffic. They formulated a network-wise utility maximization problem that aims at properly distributing the offloaded traffic among different cells. In [30], Chen et al. also considered the intercell interferences among small cells when offloading traffic, and proposed an optimal on–off control policy to manage small cells for maximizing the network-wise energy efficiency. In [31], Kang et al. considered the co-channel interference among MUs who offload data to a same small cell, and formulated an optimal user-association problem to maximize the overall network utility. In [32], Iosifidis et al. accounted for the limited network resource, and formulated a social welfare optimization problem that properly distributes the offloading traffic demand to different cells.
- **User-oriented Traffic Offloading**: The user-oriented traffic offloading focuses on investigating the users' performance gain through small-cell-based traffic offloading. Offloading to small cells can benefit the MUs in terms of reducing the mobile data cost and radio resource usage. However, due to limited and discontinuous coverage of small cells as well as nonideal backhaul, offloading might incur additional transmission delay. Thus, it is crucial to understand under which circumstance the MUs' traffic offloading is beneficial, or how the MUs can adjust the traffic offloading strategy to trade off between the benefit from offloading and the consequent downside. In [33], Cheung and Huang considered the delay in selecting the proper cells to offload data and formulated an optimization problem to trade off the benefit of reducing the MU's mobile data cost and the consequently degraded delay-performance for file transfer. In [34, 35], the authors exploited the MUs' delay-tolerance and proposed different incentive-based schemes to motivate the MUs to execute traffic offloading to small cells (to relieve traffic pressure on macrocells). In [36], Im et al. considered the MU's budget of mobile data usage, and proposed an MU-centric cost-aware traffic offloading scheme to optimize the MU's throughput-delay tradeoff.

We next illustrate two special types of small-cell-based traffic offloading, namely, the WiFi-based traffic offloading and traffic offloading with dual connectivity.

1.3.1.1 WiFi-Based Traffic Offloading

WiFi-based traffic offloading plays an indispensable role in small-cell-based traffic offloading, since more than 80% of mobile traffic nowadays are either originating or terminating in indoor circumstances (i.e., within WiFi coverage) [37]. WiFi systems, as a special tier of small cells based on the series of IEEE 802.11 standardization [6], operate on unlicensed bands. In addition to the benefits such as providing high throughput and low deployment cost, WiFi-based offloading facilitates migrating traffic from the congested licensed bands to the unlicensed ones, which thus relieves cellular operators' traffic burden on licensed bands. Many major cellular operators therefore have spent significant efforts in deploying and integrating WiFi APs into cellular systems. In [38], Lee et al., by collecting and analyzing real data, showed that WiFi systems can help offload about 65% of total mobile traffic and save 55% of mobile devices' battery power. Performance analysis (such as expected offloading delay and offloading throughput) regarding WiFi-based traffic offloading has been presented in [39, 40], which took into account issues such as the availability of WiFi APs, user's mobility, offloading strategies, and traffic intensity. As mentioned before, due to limited and discontinuous coverage of WiFi, the MUs might need to delay traffic offloading in order to wait for the next available WiFi connection. Such a delay is beneficial to the MUs, since offloading through WiFi connection provides a more favorable transmission (such as high throughput, low power consumption, and low mobile data cost). Thus, different delay-aware traffic offloading strategies have been proposed in [33, 36, 41, 42]. Accurate predictions of availability of small cells along with the MUs' traffic demands, e.g., based on collecting and analyzing historical data, will facilitate the designs of delay-aware traffic offloading strategies.

From the implementation perspective, proper geographical deployment of WiFi APs is of practical importance to the performance of WiFi-based offloading [43, 44]. Moreover, WiFi-based offloading requires a tightly integrated management of cellular systems and WiFi systems, since these two systems are separately designed in principle and might even be maintained by different operators. In particular, cellular systems and WiFi systems necessitate deep inter-network management covering core functions such as network selection, access authentication, flow control, and security, such that WiFi-based offloading can yield user-transparent flow-level mobility for simultaneously exploiting multiple different radio connections [45]. Finally, from the economic perspective, as cellular systems and WiFi systems might be owned by different network operators, different network operators need to reach a collaborative and profitable business agreement to realize a multiparty-win outcome through WiFi-based traffic offloading.

1.3.1.2   Traffic Offloading Through Small-Cell Dual Connectivity

Small-cell dual connectivity (DC) is a new paradigm proposed by industrial prac-
tices [46, 47] and 3GPP LTE-A standardizing activities [48, 49] to facilitate traffic
offloading in cellular networks. With dual connectivity, each MU can simultaneously communicate with a macrocell BS and a small-cell AP via two different radio
interfaces. The benefits of dual connectivity are two-folded, (i) it enables the separa-
tion of the control-plane and data-plane in the cellular systems, namely, macrocells
for connection and mobility managements, while small cells for data delivery, and
(ii) it yields a more efficient traffic offloading, i.e., the MUs can flexibly schedule
their traffic between macrocells and small cells. For instance, an MU can schedule its
delay-sensitive small-volume data traffic (e.g., voice-over-IP traffic) to the macrocell
BS, and meanwhile offload its delay-tolerant large-volume traffic (e.g., file down-
loading/uploading) to a small-cell AP. Proper radio resource management is crucial
to reap the benefit of traffic offloading with dual connectivity, since the MU needs
to carefully schedule its traffic and limited radio resources (e.g., power budget) for
two radio interfaces which are involved in dual connectivity. In [50], Jha et al. provided a brief survey about the key technical challenges for the small-cell-based traffic
offloading with dual connectivity and evaluated the impact of resource-splitting on
two radio interfaces. Flow control is also an important issue in traffic offloading
with dual connectivity, since the traffic delivered through the macrocell and small
cell might experience different levels of delay [51]. We will provide a more detailed
descriptions about small-cell-based traffic offloading through dual connectivity in
Chap. 2.

1.3.2   D2D-Assisted Traffic Offloading

Traffic offloading through mobile users' D2D cooperations (also referred as D2D-
assisted traffic offloading) is another important paradigm of traffic offloading in
HCNs. As we have illustrated in Sect. 1.2, D2D communication is a recently proposed paradigm in 3GPP standardization that enables two MUs of close proximity
to directly communicate with each other instead of going through BS. Thus, D2D
communications can help offload MUs' traffic from cellular links to D2D-links,
which, by exploiting MUs' close-proximity, can provide better link quality and more
efficient resource utilization [52]. The paradigm of D2D communications was first
proposed to target for proximity-based services such as local message dissemination.
Recently, D2D-assisted offloading has attracted lots of interests, since it can relieve
the heavy traffic burden on cellular links. Simulation-based studies have showed the
promising gain in reducing cellular traffic by using D2D-assisted offloading through
WiFi-direct [53], and the analytical model and performance evaluation regarding the
gain from dynamic D2D-assisted offloading were provided in [54].

D2D-assisted offloading particularly fits applications such as content distribution
for a group of neighboring MUs who are interested in receiving a common group of

contents. Instead of directly broadcasting contents to all MUs, cellular BS can first send the contents to some properly selected MUs, who then exploit D2D-links to offload the contents to others. Such a D2D-assisted offloading can achieve two types of benefits: (i) it reduces the traffic volume on cellular link, and (ii) it reduces resource usage by exploiting MUs' close proximity. However, to reap these benefits, the D2D-assisted traffic offloading requires a proper design of MUs' offloading strategies and the content transmission control. Depending on whether the content distribution is delay-sensitive, there are two categories of studies: (i) the D2D-assisted offloading for distributing real-time traffic [55–58], and (ii) the D2D-assisted offloading for distributing non-real-time traffic without a hard deadline [59–64]. We will provide more detailed discussions regarding the two groups of studies in Chap. 3.

1.4 Open Issues in Traffic Offloading in Heterogeneous Small-Cell Networks

In spite of the existing studies, there still exist many open issues regarding traffic offloading in heterogeneous small-cell networks, such as QoS-provisioning, backhaul capacity, and security. We provide a brief discussion about them in this section.

QoS-Provisioning in Traffic Offloading: The large number of small cells provide high-quality links for MUs' traffic delivery and are expected to improve the MUs' QoS. On one hand, macrocell users' QoS can be improved by exploiting traffic offloading, since traffic offloading can effectively reduce traffic burden and congestion in macrocells. On the other hand, QoS-provisioning for the offloaded users becomes more complicated. Although small cells can provide better link quality, aggressively offloading MUs' traffic to small cells may lead to negative impacts such as severe co-channel interference and excessive MUs' transmission collisions, which adversely influence the MUs' experienced QoS during offloading. Moreover, non-ideal backhauls (e.g., limited capacity, delay, and packet loss) also impair the MUs' experienced QoS. Therefore, proper management of traffic offloading/scheduling to small cells is required to guarantee the MUs' required QoS. Moreover, from the implementation perspective, advanced service level agreements (SLAs) among the operators of macrocells, small cells, and backhaul links are required to guarantee the MU's required QoS in traffic offloading. For example, dynamic path bandwidth measurement techniques are required to periodically estimate the available backhaul capacity for offloading. If the available capacity is below a threshold, then no traffic offloading is involved.

Backhaul Connection and Nonideal Property: The backhaul connection is crucial to the performance of traffic offloading, since small cells rely on backhaul for delivering the offloaded traffic to and from core networks [65]. With the demand of traffic offloading keeps increasing rapidly, backhaul capacity becomes a critical factor that limits the performance gain of traffic offloading. Different types of backhauls have different properties (e.g., capacity and latency), which need to be taken into account in traffic offloading design. For instance, the use of broadband Internet connection (such as DSL) as backhaul might incur offloading delay and packet loss due to limited DSL capacity and congestions in IP networks. In [66], Yang et al. took into account the impact of limited backhaul capacity when offloading MUs to small cells and proposed a resource allocation and traffic offloading scheme to maximize the operator's reward in traffic offloading. In [67], Samarakoon et al. took into account different cells' heterogeneous backhauls and proposed a distributed learning algorithm for properly offloading users' data to small cells without incurring to much delay. Different from assuming a fixed backhaul capacity as [66, 67], dynamically adjusting the backhaul capacity (e.g., based on the offloading needs) is a more efficient approach to exploit the benefit of the data offloading. In [68], Liu et al. proposed a joint optimization of the bandwidth allocation for wireless backhaul and the downlink power control for small cells, with the objective of maximizing the energy efficiency of all small-cell users. In [69], Wang et al. proposed a joint optimization of the backhaul capacity and the user-association in small-cell networks, with the objective of maximizing all users' total throughput.

Unlicensed Bands for Traffic Offloading: Due to limited licensed spectrum resources, wireless industries and cellular operators have been examining the feasibility of exploring unlicensed bands (such as 2.4 and 5 GHz ISM bands) to facilitate traffic offloading [70]. Exploiting unlicensed bands for traffic offloading saves the use of operators' licensed spectrums, and moreover, the availability of vast unlicensed spectrums provides high data rate and low latency for traffic offloading. As mentioned before, the key issue in exploiting unlicensed spectrums (for traffic offloading) is to achieve the harmonic coexistence with the incumbent systems which already operate on unlicensed bands. In {71], Wu et al. took into account the uncontrollable interference from incumbent users and studied the corresponding optimal resource allocation for traffic offloading that exploits unlicensed bands. In [72], Almeida et al. proposed a blank subframe allocation scheme to avoid interference/collision between LTE systems and WiFi systems. Estimating the number of co-channel transmitters and knowing the deployment density of network nodes are crucial to achieve the harmonic coexistence. In [7], Zhang et al. introduced an architecture where small cells exploit the same unlicensed spectrums as WiFi systems, and proposed an interference avoidance scheme based on estimating the density of nearby WiFi access points to facilitate harmonic coexistence between small cells and WiFi systems. In [73], Wang et al. proposed an interference estimation technique, which enables each BS to sense unlicensed spectrum and estimate the number of co-channel transmissions in a defined zone, to avoid interference between macro-BS and incumbent users in unlicensed bands.

Secrecy Issues in Traffic Offloading: With the growing concerns on security risk in wireless data services, secrecy-provisioning for traffic offloading through small cells has become an interesting issue [74, 75]. In HCNs, offloading traffic through small cells leads to several security risks. One of the most critical issues is the trustability of the small cells. Since small cells (e.g., WiFi APs) might be operated by different owners, an unauthorized and malicious AP might play as a normal AP to induce the MUs' offloading and obtain sensitive information. Another critical issue is eavesdropping. Due to the growing need for using unlicensed spectrums to support offloading (e.g., the emerging paradigm of LTE over unlicensed bands), a malicious node can exploit the open-access nature of unlicensed bands and intentionally eavesdrop the MU's offloaded traffic by receiving the MU's radio signal. As a result, if an MU blindly offloads the data, it might suffer from a risk of being eavesdropped. There are also several other important security issues, such as jamming attacks to block small cells that provide offloading service.

To address the security issues, a variety of protocol-based or application-oriented schemes such as encryption, authentication, and secure socket layer (SSL) schemes have been adopted by 3GPP standardizations [76]. Beyond these protocol-based or application-oriented secrecy-enhancing schemes, a promising and fundamental approach to address the security issue is to exploit the advanced physical layer security. The physical layer secrecy capacity, defined as the difference between the channel capacity of legitimate channel and that of the eavesdropper's channel, provides a fundamental measure of the achievable capacity for a point-to-point link which is impossible to be eavesdropped. Hence, the physical layer secrecy capacity provides an efficient manner to quantify how secure it is when an MU offloads data to the AP. In particular, it fits the needs to understand/study the secrecy-provisioning for traffic offloading, which is featured by flexible traffic scheduling and radio resource allocations. The authors of [77, 78] adopted the physical layer secrecy capacity and quantified the corresponding secrecy outage probability which depends on the MUs' offloading rate and power allocation. Furthermore, the joint traffic scheduling and power allocation schemes have been proposed to minimize the MUs' total power consumption while satisfying the MUs' required secrecy-levels.

Economics and Incentives of Traffic Offloading: The prevalence of traffic offloading in HCNs requires a good understanding of potential economic benefits. On one hand, from the MUs' perspective, traffic offloading through small cells such as WiFi APs can effectively reduce the MUs' mobile data cost. However, due to discontinuous coverage and nonideal backhaul of small cells, aggressively traffic offloading might result in performance loss (such as additional delay and packet loss). Hence, it is crucial to study how much the MUs can benefit in terms of the mobile data cost reduction, while properly trading off the degraded performance [31, 36].

From the operator's perspective, gaining sufficient economic reward from traffic offloading is a more complicated issue. First, traffic offloading to small cells involves cooperation and negotiation between macrocells and small cells. If the macrocells and small cells are owned by different operators, then it is crucial to study the revenue-sharing schemes between the macrocells and small cells to provide enough incentives to both sides. Second, densely deployed small cells owned by different operators might compete with each other to provide traffic offloading service. Hence, it is important to study the competition (e.g., the pricing strategy) among different small cells for offloading services while trading off their respective resource usages.

1.5 Aim of the Brief

The aim of the brief is to present the importance and design methodologies of the optimal radio resource allocations for traffic offloading in HCNs. To this end, we introduce two approaches corresponding to the two most important paradigms of traffic offloading in HCNs, namely, the small-cell-based traffic offloading and the D2D-assisted traffic offloading. First, regarding the paradigm of small-cell-based traffic offloading, we investigate how to jointly optimize the traffic scheduling and the corresponding radio resource allocations, with the objective of minimizing the MUs' total mobile data cost. The details will be presented in Chap. 2. Second, regarding the paradigm of D2D-assisted traffic offloading for content distribution, we study how to jointly optimize the MUs' cooperative offloading strategies and the content transmission control, with the objective of minimizing the overall radio resource usage. The details will be presented in Chap. 3. Finally, Chap. 4 concludes this brief and provides the potential future research directions.

References

1. CISCO, "Cisco visual networking index: global mobile data traffic forecast update, 2015-2020 white paper," http://www.cisco.com/c/en/us/solutions/collateral/service-provider/visual-networking-index-vni/mobile-white-paper-c11-520862.html, 2016.
2. A. Ghosh, N. Mangalvedhe, R. Ratasuk, B. Mondal, M. Cudak, E. Visotsky, T. Thomas, J. Andrews, P. Xia, H. Jo, H. Dhillon, and T. Novlan, "Heterogeneous cellular networks: From theory to practice," *IEEE Communications Magazine*, vol. 50, no. 6, pp. 54–64, 2012.
3. M. Agiwal, A. Roy, and N. Saxena, "Next generation 5G wireless networks: A comprehensive survey," *IEEE Communications Survey and Tutorials*, vol. 18, no. 3, pp. 1617–1655, 2016.
4. C. B. Networks, "Backhauling X2," http://cbnl.com/sites/all/files/userfiles/files/Backhauling-X2.pdf, 2011.
5. V. Chandrasekhar, J. Andrews, and A. Gatherer, "Femtocell networks: A survey," *IEEE Communications Magazine*, vol. 46, no. 9, pp. 59–67, 2008.
6. H. Omar, K. Abboud, N. Cheng, K. Maekshan, A. Camage, and W. Zhuang, "A survey on high efficiency wireless local area networks: next generation WiFi," *IEEE Communications Surveys & Tutorials*, to appear.
7. H. Zhang, X. Chu, W. Guo, and S. Wang, "Coexistence of WiFi and heterogeneous small cell networks sharing unlicensed spectrum," *IEEE Communications Magazine*, vol. 53, no. 3, pp. 158–164, 2015.
8. H. Yang, J. Lee, and T. Quek, "LTE-Advanced: Next-generation wireless broadband technology," *IEEE Wireless Communications*, vol. 17, no. 3, pp. 10–22, 2010.
9. X. Chu, D. Lopez-Perez, Y. Yang, and F. Gunnarsson, *Heterogeneous Cellular Networks: Theory, Simulation and Deployment*. Cambridge University Press, 2013.
10. P. Xia, V. Chandrasekhar, and J. G. Andrews, "Open vs. closed access femtocells in the uplink," *IEEE Transactions on Wireless Communications*, vol. 9, no. 12, pp. 3798–3809, 2010.
11. P. Wang, W. Song, D. Niyato, and Y. Xiao, "QoS-aware cell association in 5G heterogeneous networks with massive MIMO," *IEEE Network Magazine*, vol. 29, no. 6, pp. 76–82, 2015.
12. iPass, "iPass application," https://www.ipass.com.
13. G. Roche, A. Valcarce, D. Lopez-Perez, and J. Zhang, "Access control mechanisms for femtocells," *IEEE Communications Magazine*, vol. 48, no. 1, pp. 33–39, 2010.

14. D. Lopez-Perez, A. Valcarce, A. Ladanyi, G. Roche, and J. Zhang, "Intracell handover for inter-ference and handover mitigation in OFDMA two-tier macrocell-femtocell networks," *EURASIP Journal on Wireless Communications and Networking*, 2010, DOI:10.1155/2010/142629.

15. N. Zhang, S. Zhang, S. Wu, J. Ren, J. W. Mark, and X. Shen, "Beyond coexistence: Traffic steering in LTE networks with unlicensed bands," *to appear in IEEE Wireless Communications*, 2016.

16. K. Deppler, M. Rinne, C. Wijting, C. Ribeiro, and K. Hugl, "Device-to-Device communication as an underlay to LTE-Advanced networks," *IEEE Communications Magazine*, vol. 47, no. 12, pp. 42–49, 2009.

17. D. Feng, L. Lu, Y. Wu, G. Li, S. Li, and G. Feng, "Device-to-Device communications in cellular networks," *IEEE Communications Magazines*, vol. 52, no. 6, pp. 49–55, 2014.

18. L. Wei, R. Hu, Y. Qian, and G. Wu, "Enable Device-to-Device communications underlay-ing cellular networks: Challenges and research aspects," *IEEE Communications Magazines*, vol. 52, no. 6, pp. 90–96, 2014.

19. M. Zulhasnine, C. Huang, and A. Srinivasan, "Efficient resource allocation for Device-to-Device communication underlaying LTE network," in *Proc. of IWCMC*, Caen, France, June 2010.

20. K. Zhou, M. Wang, K. Yang, J. Zhang, W. Sheng, Q. Chen, and X. You, "Proximity discovery for Device-to-Device communications over a cellular network," *IEEE Communications Magazine*, vol. 52, no. 6, pp. 98–107, 2014.

21. K. Akkarajitsakul, P. Phunchongharn, E. Hossain, and V. Bhargava, "Energy-efficient resource sharing for mobile Device-to-Device multimedia communications," in *Proc. of IEEE ICCS*, New Orleans, LA, November 2012.

22. H. Min, W. Seo, J. Lee, S. Park, and D. Hong, "Reliability improvement using receive mode selection in the Device-to-Device uplink period underlaying cellular networks," *IEEE Transactions Wireless Communications*, vol. 10, no. 2, pp. 413–418, 2011.

23. C. Yu, K. Doppler, C. Ribeiro, and O. Tirkkonen, "Resource sharing optimization for Device-to-Device communication underlaying cellular networks," *IEEE Transactions on Wireless Communications*, vol. 10, no. 8, pp. 2752–2763, 2011.

24. M. Belleschi, G. Fodor, and A. Abrardo, "Performance analysis of a distributed resource allo-cation scheme for D2D communications," in *Proc. of GLOBECOM Workshops*, Houston, TX, Dec. 2011.

25. H. Chen, D. Wu, and Y. Cai, "Coalition formation game for green resource management in D2D communications," *IEEE Communications Letters*, vol. 18, no. 8, pp. 1395–1398, 2014.

26. Y. Wu, J. Wang, L. Qian, and R. Schober, "Optimal power control for energy efficient D2D communication and its distributed implementation," *IEEE Communications Letters*, vol. 19, no. 5, pp. 815–818, 2015.

27. Y. Wu, J. Huang, L. Qian, and R. Schober, "Energy-aware revenue optimization for cellular networks via Device-to-Device communication," in *Proc. of ICC*, London, United Kindom, June 2015.

28. X. Lin, J. Andrews, and A. Ghosh, "A comprehensive framework for Device-to-Device com-munications in cellular networks," https://arxiv.org/pdf/1305.4219v2.pdf.

29. C. Ho, D. Yuan, and S. Sun, "Data offloading in load coupled networks: A utility maximization framework," *IEEE Transactions on Wireless Communications*, vol. 13, no. 4, pp. 1912–1931, 2014.

30. X. Chen, J. Wu, Y. Cai, H. Zhang, and T. Chen, "Energy-efficiency oriented traffic offloading in wireless networks: A brief survey and a learning approach for heterogeneous cellular networks," *IEEE Journal on Selected Areas in Communications*, vol. 33, no. 4, pp. 627–640, 2015.

31. X. Kang, Y. Chia, S. Sun, and H. Chong, "Mobile data offloading through a thrid-party WiFi access point: An operator's perspective," *IEEE Transactions on Wireless Communications*, vol. 13, no. 10, pp. 5340–5351, 2014.

32. G. Iosifidis, L. Gao, J. Huang, and L. Tassiulas, "A double auction mechanism for mobile data offloading markets," *IEEE Transactions on Networking*, vol. 23, no. 5, pp. 1634–1647, 2014.

33. M. Cheung and J. Huang, "DAWN: Delay-aware WiFi offloading and network selection," *IEEE Journal on Selected Areas in Communications*, vol. 33, no. 6, pp. 1214–1223, 2015.

34. X. Zhuo, W. Gao, G. Cao, and S. Hua, "An incentive framework for cellular traffic offloading," *IEEE Transactions on Mobile Computing*, vol. 13, no. 3, pp. 541–555, 2014.

35. J. Lee, Y. Yi, S. Chong, and Y. Jin, "Economics of WiFi offloading: Trading delay for cellular capacity," *IEEE Transactions on Wireless Communications*, vol. 13, no. 3, pp. 1540–1544, 2014.

36. Y. Im, C. Wong, S. Ha, S. Sen, T. Kwon, and M. Chiang, "AMUSE: Empowering users for cost-aware offloading with throughput-daly tradeoffs," in *Proc. of IEEE INFOCOM*, Turin, Italy, April 2013.

37. ABI, "ABI research anticipates in-building mobile data traffic to grow by more than 600% by 2020," https://www.abiresearch.com/press/abi-research-anticipates-building-mobile-data-traf/.

38. K. Lee, J. Lee, Y. Yi, I. Rhee, and S. Chong, "Mobile data offloading: How much can WiFi deiver," *IEEE/ACM Transactions on Networking*, vol. 21, no. 2, pp. 536–550, 2013.

39. F. Mehmeti and T. Spyropoulos, "Performance analysis of "on-the-spot" mobile data offloading," in *Proc. of GLOBECOM*, Atlanta, GA, Dec. 2013.

40. D. Suh, H. Ko, and S. Pack, "Efficiency analysis of WiFi offloading techniques," *IEEE Transactions on Vehicular Technology*, vol. 65, no. 5, pp. 3813–3817, 2016.

41. M. Cheung and J. Huang, "Optimal delayed WiFi offloading," in *Proc. of International Symposion on Modeling and Optimization in Mobile, Ad Hoc and Wireless Networks(WiOpt)*, Tsukuba Science City, Japan, May 2013.

42. N. Cheng, N. Lu, N. Zhang, X. Zhang, X. Shen, and J. Mark, "Opportunistic WiFi offloading in vehicular environment: A game-theory approach," *IEEE Transations on Intelligent Transportation Systems*, vol. 17, no. 7, pp. 1944–1955, 2016.

43. L. Hu, C. Coletti, N. Huan, I. Kovacs, B. Vejlgaard, R. Irmer, and N. Scully, "Realistic indoor WiFi and femto deployment study as the offloading solution to LTE macro networks," in *Proc. of VTC-Fall*, Quebec City, Canada, Sep. 2012.

44. N. Ristanovic, J. L. Boudec, A. Chaintreau, and V. Erramilli, "Energy efficient offloading of 3G networks," in *Proc. of IEEE International Conference on MASS*, Valencia, Span, Oct. 2011.

45. "3rd generation partnership project (3GPP), 3GPP TS 24.312: Access network discovery and selection function (ANDSF) management object (MO) (Rel. 10)," Sophia Antipolis, France.

46. Qualcomm, "LTE-Advanced evolving and expanding into new frontiers," https://www.qualcomm.com/documents/lte-advanced-evolving-and-expanding-new-frontiers.

47. N. S. Networks, "LTE Release 12 and Beyond," http://resources.alcatel-lucent.com/asset/200174.

48. N. Ali, A. Taha, and H. Hassanein, "Quality of Service in 3GPP R12 LTE-Advanced," *IEEE Communications Magazine*, vol. 51, no. 8, pp. 103–109, 2013.

49. C. Sankaran, "Data offloading techniques in 3GPP Rel-10 networks: A tutorial," *IEEE Communications Magazine*, vol. 50, no. 6, pp. 46–53, 2012.

50. S. Jha, K. Sivanesan, R. Vannithamby, and A. Koc, "Dual connectivity in LTE small cell networks," in *Proc. of IEEE GLOBECOM Workshops*, Austin, TX, Dec. 2014.

51. H. Wang, C. Rosa, and K. Pedersen, "Dual connectivity for LTE-Advanced heterogeneous networks," *Wireless Networks*, vol. 22, no. 4, pp. 1315–1328, 2016.

52. J. Liu, Y. Kawamoto, H. Nishiyama, N. Kato, and N. Kadowaki, "Device-to-Device communications achieve efficient load balancing in LTE-Advanced networks," *IEEE Wireless Communications*, vol. 21, no. 2, pp. 57–65, 2014.

53. A. Pyattaev, K. Johnsson, S. Andreev, and Y. Koucheryavy, "3GPP LTE traffic offloading onto WiFi direct," in *Proc. of WCNCW*, Shanghai, China, Apr. 2013.

54. S. Andreev, O. Galinina, A. Pyattaev, K. Johnsson, and Y. Koucheryavy, "Analyzing assisted offloading of cellular user sessions onto D2D links in unlicensed bands," *IEEE Journal on Selected Areas in Communications*, vol. 33, no. 1, pp. 67–80, 2015.

55. L. Al-Kanj, Z. Dawy, W. Saad, and E. Kutanoglu, "Energy-aware cooperative content distribution over wireless networks: Optimized and distributed approaches," *IEEE Transactions on Vehicular Technology*, vol. 62, no. 8, pp. 3828–3847, 2013.

56. L. Al-Kanj, V. Poor, and Z. Dawy, "Optimal cellular offloading via Device-to-Device communication networks with fairness constraints," *IEEE Transactions on Wireless Communications*, vol. 13, no. 8, pp. 4628–4643, 2014.

57. T. Wang, L. Song, Z. Han, and B. Jiao, "Dynamic popular content distribution in vehicular networks using coalition formation games," *IEEE Journal on Selected Areas Communications*, vol. 31, no. 9, pp. 538–547, 2013.

58. N. Cheng, N. Lu, N. Zhang, X. Shen, and J. Mark, "Vehicular WiFi offloading: Challenges and solutions," *Vehicular Communications (Elsevier)*, vol. 1, no. 1, pp. 13–21, 2014.

59. Y. Li, Y. Jiang, D. Jin, L. Su, L. Zeng, and D. Wu, "Energy-efficient optimal opportunistic forwarding for delay-tolerant networks," *IEEE Transactions Vehicular Technology*, vol. 59, no. 9, pp. 4500–4512, 2010.

60. X. Wang, M. Chen, Z. Han, T. Kwon, and Y. Choi, "Content dissemination by pushing and sharing in mobile cellular networks: An analytical study," in *Proc. of IEEE MASS'2012*, Las Vegas, NV, Oct. 2012.

61. J. Whitbeck, M. Amorim, Y. Lopez, J. Leguay, and V. Conan, "Relieving the wireless infrastructure: When opportunistic networks meet guaranteed delays," in *Proc. of IEEE WoWMoM*, Lucca, Italy, Jun. 2011.

62. N. Golrezaei, A. Molisch, A. Dimakis, and G. Caire, "Femtocaching and Device-to-Device collaboration: A new architecture for wireless video distribution," *IEEE Communs. Magazine*, vol. 51, no. 4, pp. 142–149, 2013.

63. C. X. Mavromoustakis, C. Dimitriou, G. Mastorakis, and E. Pallis, "Real-time performance evaluation of F-BTD scheme for optimized QoS energy conservation in wireless devices," in *Proc. of IEEE GLOBECOM Workshops*, Atlanta, GA, Dec. 2013.

64. C. X. Mavromoustakis, G. Mastorakis, A. Bourdena, E. Pallis, G. Kormentzas, and J. Rodrigues, "Context-oriented opportunistic cloud offload processing for energy conservation in wireless devices," in *Proc. of IEEE GLOBECOM Workshops*, Austin, Texas, Dec. 2014.

65. M. Jaber, M. Imran, R. Tafazolli, and A. Tukmanov, "5G backhaul challenges and emerging research directions: A survey," *IEEE Access*, vol. 4, pp. 1743–1766, 2016.

66. Y. Yang, T. Q. S. Quek, and L. Duan, "Backhaul-constrained small cell networks: Refunding and QoS provisioning," *IEEE Transactions on Wireless Communications*, vol. 13, no. 9, pp. 5148–5161, 2014.

67. S. Samarakoon, M. Bennis, W. Saad, and M. Latva-aho, "Backhaul-aware interference management in the uplink of wireless small cell networks," *IEEE Transactions on Wireless Communications*, vol. 12, no. 11, pp. 5813–5825, 2013.

68. H. Liu, H. Zhang, J. Cheng, and V. C. Leungz, "Energy efficient power allocation and backhaul design in heterogeneous small cell networks," *IEEE Transactions on Wireless Communications*, vol. 12, no. 11, pp. 5813–5825, 2013.

69. N. Wang, E. Hossain, and V. K. Bhargava, "Joint downlink cell association and bandwidth allocation for wireless backhauling in two-tier hetnets with large-scale antenna arrays," *IEEE Transactions on Wireless Communications*, vol. 15, no. 5, pp. 3251–3268, 2016.

70. Qualcomm, "Best use of unlicensed spectrum for 1000x," https://www.qualcomm.com/media/documents/files/whitepaper-making-the-best-use-of-unlicensed-spectrum.pdf.

71. Y. Wu, H. Chai, L. Qian, W. Lu, Q. Zhao, and C. Yu, "Energy-aware optimal data offloading over unlicensed spectrums," in *Proc. of IEEE 84th Vehicular Technology Conference: VTC2016-Fall*, Montreal, Canada, Sep. 2016.

72. E. Almeida, A. M. Cavalcante, R. C. D. Paiva, F. S. Chaves, F. M. A. Jr, and R. D. Vieira, "Enabling LTE/WiFi coexistence by LTE blank subframe allocation," in *Proc. of IEEE International Conference on Communications*, Budapest, Hungary, June 2013.

73. S. Wang, W. Guo, and M. D. McDonnell, "Downlink interference estimation without feedback for heterogeneous network interference avoidance," in *Proc. of 21st International Conference on Telecommunications*, Lisbon, Portugal, May 2014.

74. ERICSSON, "5G security: Scenarios and solutions," http://www.ericsson.com/res/docs/whitepapers/wp-5g-security.pdf.

75. HUAWEI, "5G security: Forward thinking Huawei white paper," http://www.huawei.com/minisite/5g/img/5G_Security_Whitepaper_en.pdf, 2015.

76. N. Yang, L. Wang, G. Geraci, M. Elkashlan, and J. Yuan, "Safeguarding 5G wireless communication networks using physical layer security," *IEEE Communications Magazine*, vol. 53, no. 4, pp. 20–27, 2015.

77. Y. Wu, K. Guo, J. Huang, and X. Shen, "Secrecy-based energy-efficient data offloading via dual-connectivity over unlicensed spectrums," *IEEE Journal on Selected Areas in Communications*, vol. 34, no. 12, pp. 1–19, 2016.

78. Y. Wu, J. Zheng, K. Guo, L. Qian, X. Shen, and Y. Cai, "Secrecy guaranteed optimal traffic offloading via dual-connectivity in small cell networks," in *Proc. of IEEE WCSP*, Yangzhou, China, Oct. 2016.

Chapter 2
Resource Allocation for Small-Cell-Based Traffic Offloading

Traffic offloading through small cells is an efficient approach to address the rapidly growing traffic demand in cellular systems. To facilitate traffic offloading, the recent 3GPP Release 12 has proposed a new paradigm of small-cell dual connectivity (DC) that allows a mobile user (MU) to simultaneously communicate with a macro base station (BS) and a small-cell access point (AP) through two different radio interfaces [1, 2]. With DC, an MU can flexibly schedule its traffic to the BS and offload traffic to small-cell AP simultaneously, hence achieving the benefits, such as reducing mobile data cost and improving radio resource utilization. However, in order to achieve these benefits of traffic offloading, we need to properly design the radio resource allocations due to the MUs' limited radio resources.

In this chapter, we consider the paradigm of the small-cell-based MUs' uplink traffic offloading with DC. Regarding this paradigm, we investigate the MUs' joint traffic scheduling and power allocation problem, with the objective of minimizing the MUs' overall mobile data cost through properly offloading traffic to a small cell, while avoiding excessive co-channel interference when offloading. Such a problem is motivated by the following considerations. Offloading traffic to small cell can save the MUs' mobile data cost, since the small cell usually charges the MUs' traffic at a lower price than the macrocell. However, severe co-channel interferences will be incurred if all MUs aggressively offload traffic to the small cell without coordination. The severe co-channel interferences consume the MUs larger transmit-powers, which adversely compromise the benefit of traffic offloading. Therefore, we formulate the optimization problem that jointly determines the MUs' traffic scheduling and power allocations to the BS and AP. Our objective is to minimize the MUs' total data cost, while satisfying each MU's transmit-power constraint through proper interference control. Our contributions in this chapter are summarized as follows.

© The Author(s) 2017
Y. Wu et al., *Radio Resource Management for Mobile Traffic Offloading in Heterogeneous Cellular Networks*, SpringerBriefs in Electrical and Computer Engineering, DOI 10.1007/978-3-319-51037-8_2

- **Optimization Formulation**: We formulate a cost minimization problem for a group of MUs' uplink traffic offloading with DC, in which each MU jointly determines its traffic scheduling and the transmit-power allocation to the AP and BS. The objective is to minimize all MUs' total mobile data cost, while meeting each MU's traffic demand and its transmit-power constraint.

- **Algorithm Design**: We propose a centralized algorithm to solve the joint traffic scheduling and power allocation problem. Despite the nature of nonconvexity of the problem, we perform a series of manipulations to transform the joint optimization problem into an equivalent SINR-assignment problem (here, SINR refers to each MU's received signal-to-interference-plus-noise ratio at the AP). We then explore the hidden monotonicity of the SINR-assignment problem, and design a two-layered algorithm to solve it and obtain the optimal SINRs at the AP. Using the obtained optimal SINRs, we finally derive the MUs' optimal traffic scheduling and transmit-powers to minimize all MUs' total data cost.

- **Performance Evaluation**: Numerical results show that the proposed algorithm can achieve the close-to-optimum results which are obtained by the LINGO (a commercial optimization software [3]), but using a significantly less computational time. The numerical results also validate that the optimized traffic offloading can significantly reduce the MUs' data cost. Specifically, the optimized traffic offloading can save more than 75% of the total cost compared with the scheme of no offloading, and save more than 65% of the total cost compared with a fixed offloading scheme.

The remainder of this chapter is organized as follows. Section 2.1 provides a review of related studies. We present the system model and the problem formulation in Sect. 2.2. In Sect. 2.3, we present a series of transformations that transform the original joint optimization problem into an equivalent SINR-assignment problem. We present an efficient algorithm to solve the SINR-assignment problem in Sect. 2.4. We next show the numerical results in Sect. 2.5 and discuss about the possible extension in Sect. 2.6. Finally, we summarize the chapter in Sect. 2.7. The complete result of this chapter (including all technical proofs) can be found in [4].

2.1 Related Studies

We review two groups of studies which are close to the work in this chapter, namely, those about the small-cell-based traffic offloading with DC and those about the small-cell-based traffic offloading taking into account the co-channel interference.

(1) Traffic offloading with DC: In [5], Jha et al. provided a brief survey about the key challenges of the small-cell dual connectivity. In [6], Mukherjee et al. proposed a pairing scheme that facilitates macrocell BSs and small-cell APs to form DCs with the MUs for executing traffic offloading. Considering the MU's limited transmit-power capacity to execute DC, Liu et al. in [7] proposed an enhanced uplink power control scheme that splits the MU's transmit-power budget to the macrocell and small cells.

In [8, 9], Mukherjee, and Wang et al. studied the flow control schemes for traffic offloading via DC while taking into account the backhaul capacity constraint.

(2) Traffic offloading taking into account the co-channel interference: An important issue regarding traffic offloading is how to properly manage the co-channel interference among the MUs when offloading traffic to the same AP. Related studies about this issue can be further categorized into the following two subgroups.

- **Interference in downlink traffic offloading**: In [10], Zhang et al. considered the case of downlink interference and proposed a power allocation scheme for access providers (which share a common spectrum) to enhance their downlink offloading capacities. In [11], Wang et al. exploited the MUs' device-to-device cooperation for data offloading and considered the MUs' interference due to sharing the same channel. In [12], Ye et al. proposed a scheme for scheduling the MUs' traffic to different small cells, while taking into account the inter-cell interference. In [13], Ho et al. investigated the load-coupling effect due to inter-cell interference and proposed a traffic allocation to distribute traffic between macrocells and small cells. In [14], Chen et al. investigated the time-varying traffic and proposed a dynamic decision model to compute the strategy for offloading traffic from macrocells to small cells. In [15], Iosifidis et al. proposed an auction mechanism that facilitates an efficient data offloading from mobile network operators to small-cell APs while taking into account the interference among the APs.
- **Interference in uplink traffic offloading**: There also exist studies considering the co-channel interference in the MUs' uplink traffic offloading. In [16], Yang et al. considered the MUs' co-channel interference when the MUs are offloaded to the AP and designed a refunding scheme to incentivize the privately owned APs to admit the offloaded MUs. In [17], Kang et al. accounted for the MUs' co-channel interference when offloading traffic to the AP, and designed access-selection schemes for maximizing the utility of network operator. In [18], Wu et al. proposed a joint cell-association and power allocation scheme that takes into account the MUs' co-channel interferences when offloading data.

2.2 System Model and Problem Formulation

2.2.1 System Model

Figure 2.1 illustrates the system model considered in this chapter. The model is comprised of one BS, one AP, and a group of MUs $\mathscr{I} = \{1, 2, ..., I\}$. With DC, the MUs perform uplink traffic delivery to the BS and AP. Each MU i has two radio-interfaces, one for sending traffic to the BS and another for offloading traffic to the AP. We use x_{iA} and x_{iB} to denote MU i's transmission rates to the AP and BS, respectively, and we use p_{iA} and p_{iB} to denote MU i's transmit-powers to the AP and

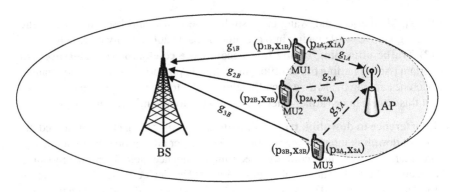

Fig. 2.1 Illustration of the system model comprised by one BS, one AP and a group of three MUs ($\mathscr{I} = \{1, 2, 3\}$)

BS, respectively. The subscripts "A" and "B" denote "AP" and "BS," respectively. We consider that the BS and AP operate on different channels.

The small-cell AP allows multiple MUs to share its channel for uplink traffic offloading [19, 20], which saves spectrum usage but leads to co-channel interference among the MUs when offloading traffic. To make the discussions more concrete, we adopt the throughput model as [14, 16, 17], i.e., given the MUs' transmit-powers $\{p_{iA}\}_{i \in \mathscr{I}}$, MU i's transmission rate to the AP is

$$x_{iA} = W \log_2 \left(1 + \frac{p_{iA} g_{iA}}{\sum_{j \neq i, j \in \mathscr{I}} p_{jA} g_{jA} + n_A} \right), \forall i \in \mathscr{I}, \qquad (2.1)$$

where W denotes the AP's channel bandwidth, and g_{iA} denotes the channel gain from MU i to the AP. Notation $n_A = W n_0$ denotes the power of the background noise at the AP, with n_0 denoting the power density.

The BS allocates orthogonal subchannels to different MUs for accommodating the uplink traffic (such as in Orthogonal Frequency Division Multiple Access, OFDMA). Given MU i's transmit-power p_{iB}, MU i's uplink transmission rate to the BS is

$$x_{iB} = B \log_2 \left(1 + \frac{p_{iB} g_{iB}}{n_B} \right), \forall i \in \mathscr{I}, \qquad (2.2)$$

where g_{iB} is the channel power gain from MU i to the BS, and B is the bandwidth allocated to MU i by the BS (we consider that the BS's channel allocation is given in this study). Notation $n_B = B n_0$ denotes the power of the background noise at the BS.

Each MU i requires to achieve the targeted traffic rate R_i^{req} with DC, i.e.,

$$x_{iA} + x_{iB} \geq R_i^{\text{req}}, \forall i \in \mathscr{I}. \qquad (2.3)$$

Table 2.1 Key notations used in this chapter

p_{iA}	MU i's transmit-power to the AP	$n_A = W n_0$	Background noise power at the AP
p_{iB}	MU i's transmit-power to the BS	$n_B = B n_0$	Background noise power at the BS
x_{iA}	MU i's throughput to the AP	R_i^{req}	MU i's traffic demand
x_{iB}	MU i's throughput to the BS	π_A	Unit data price announced by the AP
g_{iA}	Channel power gain from MU i to the AP	π_B	Unit data price announced by the BS
g_{iB}	Channel power gain from MU i to the BS	P_{iA}^{max}	MU i's maximum transmit-power to the AP
W	AP's channel bandwidth	P_{iB}^{max}	MU i's maximum transmit-power to the BS
B	BS's channel bandwidth	P_i^{max}	MU i's maximum total transmit-power
n_0	Power density of background noise		

Finally, we consider the usage-based pricing scheme by both the AP and BS [16, 17, 21]. Let π_A and π_B denote the unit-prices announced by the AP and BS, respectively. MU i's mobile data cost in one unit time is given by:

$$C_i(x_{iA}, x_{iB}) = \pi_A x_{iA} + \pi_B x_{iB}, \forall i \in \mathscr{I}. \tag{2.4}$$

Table 2.1 summarizes the key notations used in this chapter.

2.2.2 Problem Formulation

To minimize the MUs' total mobile data cost while controlling their co-channel interference when offloading data to the AP, we formulate an optimization problem that jointly optimizes the MUs' traffic scheduling $\{x_{iA}, x_{iB}\}_{i \in \mathscr{I}}$ and the transmit-powers $\{p_{iA}, p_{iB}\}_{i \in \mathscr{I}}$ as follows (Here, "CMP" stand for "Cost Minimization Problem"):

$$\text{(CMP): Minimize} \sum_{i \in \mathscr{I}} C_i(x_{iA}, x_{iB}) = \sum_{i \in \mathscr{I}} \pi_A x_{iA} + \sum_{i \in \mathscr{I}} \pi_B x_{iB}$$

$$\text{Subject to: } 0 \leq p_{iA} \leq P_{iA}^{max}, \forall i \in \mathscr{I}, \tag{2.5}$$

$$0 \leq p_{iB} \leq P_{iB}^{max}, \forall i \in \mathscr{I}, \tag{2.6}$$

$$p_{iA} + p_{iB} \leq P_i^{max}, \forall i \in \mathscr{I}, \tag{2.7}$$

Constraints (2.1), (2.2), and (2.3),

Variables: (x_{iA}, x_{iB}), $\forall i \in \mathscr{I}$ and (p_{iA}, p_{iB}), $\forall i \in \mathscr{I}$.

Constraints (2.5) and (2.6) mean that MU i's transmit-power to the AP and BS cannot exceed the respective power capacities denoted by $P_{iA}^{\max}$ and $P_{iB}^{\max}$. Constraint (2.7) ensures that MU i's total power consumption for the two interfaces cannot exceed the upper bound $P_i^{\max}$. Problem (CMP) is a nonconvex optimization problem which is difficult to solve in general. We develop efficient algorithm to solve Problem (CMP) and derive the MUs' optimal offloading solution. In this chapter, to focus on deriving the optimal offloading solution and showing the consequent benefits, we assume that Problem (CMP) is always feasible.

2.3 Equivalent Problem Transformations

In this section, we present a series of transformations of Problem (CMP) into an equivalent form. By exploiting the hidden layered-structure of this equivalent form, we design efficient algorithms to solve it and obtain the optimal offloading solution in the next section.

We first identify the following property of Problem (CMP) (all the detailed proofs can be referred in [4]).

Lemma 2.1 *Constraint (2.3) is tight at any optimal solution of Problem (CMP).*

By using Lemma 2.1 and Eq. (2.2), we express x_{iB} and p_{iB} as functions of p_{iA} for each MU i as:

$$x_{iB} = R_i^{\text{req}} - W \log_2 \left(1 + \frac{p_{iA} g_{iA}}{\sum_{j \neq i, j \in \mathscr{I}} p_{jA} g_{jA} + n_A} \right), \forall i \in \mathscr{I}, \quad (2.8)$$

$$p_{iB} = \frac{n_B}{g_{iB}} 2^{\frac{R_i^{\text{req}}}{B}} \frac{1}{(1 + \frac{p_{iA} g_{iA}}{\sum_{j \neq i, j \in \mathscr{I}} p_{jA} g_{jA} + n_A})^{\frac{W}{B}}} - \frac{n_B}{g_{iB}}, \forall i \in \mathscr{I}. \quad (2.9)$$

Similar to [16, 17, 22], we consider the practical scenario that the price of the AP is lower than that of the BS (i.e., $\pi_A < \pi_B$), which motivates the MUs to offload traffic to the AP. By using (2.1), (2.8), and (2.9), we transform Problem (CMP) into an equivalent Transmit-Power Allocation Problem (TPA-P):

(TPA-P): Maximize $\sum_{i \in \mathscr{I}} (\pi_B - \pi_A) W \log_2 \left(1 + \frac{p_{iA} g_{iA}}{\sum_{j \neq i, j \in \mathscr{I}} p_{jA} g_{jA} + n_A} \right)$

Subject to: $W \log_2 \left(1 + \frac{p_{iA} g_{iA}}{\sum_{j \neq i, j \in \mathscr{I}} p_{jA} g_{jA} + n_A} \right) \leq R_i^{\text{req}}, \forall i \in \mathscr{I}, \quad (2.10)$

$p_{iA} \leq P_{iA}^{\max}, \forall i \in \mathscr{I}, \quad (2.11)$

$\frac{n_B}{g_{iB}} 2^{\frac{R_i^{\text{req}}}{B}} \frac{1}{(1 + \frac{p_{iA} g_{iA}}{\sum_{j \neq i, j \in \mathscr{I}} p_{jA} g_{jA} + n_A})^{\frac{W}{B}}} \leq P_{iB}^{\max} + \frac{n_B}{g_{iB}}, \forall i \in \mathscr{I}, \quad (2.12)$

$$\frac{n_B}{g_{iB}} 2^{\frac{R_i^{req}}{B}} \frac{1}{(1 + \frac{p_{iA}g_{iA}}{\sum_{j\neq i, j\in\mathscr{I}} p_{jA}g_{jA}+n_A})^{\frac{W}{B}}} + p_{iA} \leq P_i^{max} + \frac{n_B}{g_{iB}}, \forall i \in \mathscr{I},$$

$$\text{Variables: } p_{iA}, \forall i \in \mathscr{I}. \tag{2.13}$$

Problem (TPA-P) only involves $\{p_{iA}\}_{i\in\mathscr{I}}$ as variables. Here, (2.10) comes from Lemma 2.1, i.e., each MU i's x_{iA} cannot exceed R_i^{req}. Constraints (2.11), (2.12), and (2.13) come from (2.5), (2.6), and (2.7), respectively. Problem (TPA-P) indicates a tradeoff in offloading traffic as follows. To minimize the MUs' total cost, all MUs should offload their traffic to the AP as much as possible, since the AP charges at a lower price than the BS. However, aggressively offloading data to the AP results in a heavy co-channel interference among the MUs, which increases the MUs' transmit-powers to meet their traffic scheduling. Recall that the MUs' power allocations need to satisfy (2.11)–(2.13).

We emphasize that although Problem (TPA-P) looks similar to a variant of conventional power control problems over interference channels (e.g., [23, 24]), it is in fact much more complicated due to taking into account each MU's transmit-powers at two different radio-interfaces. This consequently yields the non-convex constraint (2.13) which couples each MU i's p_{iA} and p_{iB} together. Please also note that all MUs' $\{p_{iA}\}_{i\in\mathscr{I}}$ are also coupled due to the interference at the AP. Hence, Problem (TPA-P) is very difficult to solve.

To solve Problem (TPA-P), we need to make some further equivalent transformations. We use θ_i to denote MU i's achieved SINR at the AP as follows:

$$\theta_i = \frac{p_{iA}g_{iA}}{\sum_{j\neq i, j\in\mathscr{I}} p_{jA}g_{jA} + n_A}, \forall i \in \mathscr{I}. \tag{2.14}$$

Based on (2.14), we have the following result that connects $\{\theta_i\}_{i\in\mathscr{I}}$ with $\{p_{iA}\}_{i\in\mathscr{I}}$.

Proposition 2.1 *Given any profile of transmit-powers $\{p_{iA}\}_{i\in\mathscr{I}}$ which is feasible for Problem (TPA-P), the corresponding profile of $\{\theta_i\}_{i\in\mathscr{I}}$ given by (2.14) ensures that the following result holds:*

$$p_{iA} = \frac{n_A}{g_{iA}} \frac{\theta_i}{1+\theta_i} \frac{1}{1 - \sum_{i\in\mathscr{I}} \frac{\theta_i}{1+\theta_i}}, \forall i \in \mathscr{I}, \tag{2.15}$$

and we always have $\sum_{i\in\mathscr{I}} \frac{\theta_i}{1+\theta_i} < 1$.

Based on Proposition 2.1, we can use $\{\theta_i\}_{i\in\mathscr{I}}$ to substitute $\{p_{iA}\}_{i\in\mathscr{I}}$ and transform Problem (TPA-P) into an equivalent SINR-assignment problem as follows:

(SINR-P): Maximize $\sum_{i\in\mathscr{I}} (\pi_B - \pi_A) W \log_2 (1 + \theta_i)$

$$\text{Subject to: } 0 \leq \theta_i \leq 2^{\frac{R_i^{req}}{W}} - 1, \forall i \in \mathscr{I}, \tag{2.16}$$

$$\frac{n_A}{g_{iA}} \frac{\theta_i}{1+\theta_i} \frac{1}{1 - \sum_{i\in\mathscr{I}} \frac{\theta_i}{1+\theta_i}} \leq P_{iA}^{max}, \forall i \in \mathscr{I}, \tag{2.17}$$

$$\frac{n_B}{g_{iB}} 2^{\frac{R_i^{req}}{B}} \frac{1}{(1+\theta_i)^{\frac{W}{B}}} \leq P_{iB}^{max} + \frac{n_B}{g_{iB}}, \forall i \in \mathscr{I}, \qquad (2.18)$$

$$\frac{n_A}{g_{iA}} \frac{\theta_i}{1+\theta_i} \frac{1}{1 - \sum_{i \in \mathscr{I}} \frac{\theta_i}{1+\theta_i}} + \frac{n_B}{g_{iB}} 2^{\frac{R_i^{req}}{B}} \frac{1}{(1+\theta_i)^{\frac{W}{B}}} \leq P_i^{max} + \frac{n_B}{g_{iB}}, \forall i \in \mathscr{I}(2.19)$$

$$\sum_{i \in \mathscr{I}} \frac{\theta_i}{\theta_i + 1} < 1, \qquad (2.20)$$

Variables: $\theta_i, \forall i \in \mathscr{I}$.

Notice that we introduce constraint (2.20) in Problem (SINR-P) for the convenience of later discussions, and Proposition 2.1 shows that introducing such a constraint does not reduce the feasible set of Problem (TPA-P). Once we solve Problem (SINR-P), then the optimal solution $\{p_{iA}^*\}_{i \in \mathscr{I}}$ of Problem (TPA-P) can be computed based on Eq. (2.15).

Again due to non-convexity, Problem (SINR-P) is still very difficult to solve. To solve it efficiently, we further introduce a one-to-one mapping as follows:

$$\rho_i = \frac{\theta_i}{1+\theta_i}, \Longleftrightarrow \theta_i = \frac{\rho_i}{1-\rho_i}, \forall i \in \mathscr{I}. \qquad (2.21)$$

Since $\sum_{i \in \mathscr{I}} \rho_i < 1$ always holds (i.e., Proposition 2.1), we introduce another positive variable ρ_0 such that $\rho_0 + \sum_{i \in \mathscr{I}} \rho_i = 1$. Such a condition will help simplify the complicated denominators in (2.17) and (2.19).

By using $\{\rho_i\}_{i \in \mathscr{I}}$ and ρ_0 and some equivalent transformations, we equivalently transform Problem (SINR-P) into Problem (SINR-M-P) as follows (the letter "M" means "Medium").

(SINR-M-P):   Minimize $\sum_{i \in \mathscr{I}} (\pi_B - \pi_A) W \log_2 \left(\rho_0 + \sum_{j \neq i, j \in \mathscr{I}} \rho_j \right)$ \qquad (2.22)

Subject to: $0 \leq \rho_i \leq 1 - \frac{1}{2^{\frac{R_i^{req}}{W}}}, \forall i \in \mathscr{I}, \qquad (2.23)$

$$\frac{n_A}{g_{iA}} \frac{\rho_i}{\rho_0} \leq P_{iA}^{max}, \forall i \in \mathscr{I}, \qquad (2.24)$$

$$\rho_0 + \sum_{j \neq i, j \in \mathscr{I}} \rho_j \leq \left(\frac{P_{iB}^{max} + \frac{n_B}{g_{iB}}}{\frac{n_B}{g_{iB}} 2^{\frac{R_i^{req}}{B}}} \right)^{\frac{B}{W}}, \forall i \in \mathscr{I}, \qquad (2.25)$$

$$\frac{n_B}{g_{iB}} 2^{\frac{R_i^{req}}{B}} \left(\rho_0 + \sum_{j \neq i, j \in \mathscr{I}} \rho_j \right)^{\frac{W}{B}} + \frac{n_A}{g_{iA}} \frac{\rho_i}{\rho_0} \leq P_i^{max} + \frac{n_B}{g_{iB}}, \forall i \in \mathscr{I}, (2.26)$$

$$\rho_0 + \sum_{i \in \mathscr{I}} \rho_i = 1, \qquad (2.27)$$

Variables: ρ_0 and $\rho_i, \forall i \in \mathscr{I}$.

Fig. 2.2 Relationship between different problem formulations. We mark out the decision variables of each formulation

Notice that constraints (2.23)–(2.27) correspond to (2.16)–(2.20) in Problem (SINR-P), respectively. We can prove that the feasible region of Problem (SINR-P) and that of Problem (SINR-M-P) form an one-to-one mapping (when Problem (SINR-P) is feasible). We therefore have the following important result that connect Problem (SINR-P) and Problem (SINR-M-P).

Proposition 2.2 *Let $(\rho_0^*, \{\rho_i^*\}_{i \in \mathscr{I}})$ denote an optimal solution of Problem (SINR-M-P). Then, $\theta_i^* = \frac{\rho_i^*}{1-\rho_i^*}, \forall i \in \mathscr{I}$ corresponds to an optimal solution of Problem (SINR-P).*

Proposition 2.2 allows us to solve Problem (SINR-P) by solving Problem (SINR-M-P). A keen observation on Problem (SINR-M-P) is that it has a hidden layered structure which facilitates us to solve it. Specifically, supposing that $\rho_0 \in (0, 1]$ is given, the resulting subproblem of Problem (SINR-M-P) is a monotonic optimization problem [25, 26]. Therefore, by executing a line-search over $\rho_0 \in (0, 1]$ on the top and solving a series of monotonic subproblems, we can eventually solve Problem (SINR-M-P). This rationale leads to the following proposed algorithm to solve Problem (SINR-M-P) presented in the next section.

Before leaving this section, we present Fig. 2.2 that shows how we transform Problem (CMP) into Problem (SINR-M-P) step-by-step.

2.4 Efficient Algorithm for Optimal Offloading Solution

This section focuses on proposing an algorithm to solve Problem (SINR-M-P) by exploiting its layered structure. Recall that after obtaining the optimal solution of Problem (SINR-M-P), we can further derive the optimal offloading solution for the original Problem (CMP).

2.4.1 Layered Structure

The keen observation is that Problem (SINR-M-P) has a layered structure, i.e., the subproblem to optimize $\rho_i, \forall i \in \mathscr{I}$ under given ρ_0 in advance, and the top-problem to optimize $\rho_0 \in (0, 1]$ to minimize the original objective function. We illustrate the details about the two problems as follows.

Subproblem to optimize $\{\rho_i\}_{i \in \mathscr{I}}$ **under a given** $\rho_0 \in (0, 1]$:

$$\text{(SINR-M-SubP): } F_{\text{sub}}(\rho_0) = \text{ Minimize } \sum_{i \in \mathscr{I}} (\pi_B - \pi_A) W \log_2 \left(\rho_0 + \sum_{j \in \mathscr{I}, j \neq i} \rho_j \right) \quad (2.28)$$

$$\text{Subject to: } \{\rho_i\}_{i \in \mathscr{I}} \in \mathscr{G}_{(\rho_0)} \cap \mathscr{H}_{(\rho_0)}, \quad (2.29)$$

$$\text{Variables: } \rho_i, \forall i \in \mathscr{I}.$$

Constraint (2.29) means that the profile $\{\rho_i\}_{i \in \mathscr{I}}$ for all MUs should belong to the intersection of sets $\mathscr{G}_{(\rho_0)}$ and $\mathscr{H}_{(\rho_0)}$. Here, set $\mathscr{G}_{(\rho_0)}$ can be characterized by (2.23)–(2.26) under a given ρ_0 as follows:

$$\mathscr{G}_{(\rho_0)} = \left\{ \{\rho_i\}_{i \in \mathscr{I}} \,\middle|\, 0 \leq \rho_i \leq 1 - \frac{1}{2^{\frac{R_i^{\text{req}}}{W}}}, \forall i \in \mathscr{I}; \ \frac{n_A}{g_{iA}} \frac{\rho_i}{\rho_0} \leq P_{iA}^{\max}, \forall i \in \mathscr{I}; \right.$$

$$\sum_{j \neq i, j \in \mathscr{I}} \rho_j \leq \left(\frac{P_{iB}^{\max} + \frac{n_B}{g_{iB}}}{\frac{n_B}{g_{iB}} 2^{\frac{R_i^{\text{req}}}{B}}} \right)^{\frac{B}{W}} - \rho_0, \forall i \in \mathscr{I};$$

$$\left. \frac{n_B}{g_{iB}} 2^{\frac{R_i^{\text{req}}}{B}} \left(\rho_0 + \sum_{j \neq i, j \in \mathscr{I}} \rho_j \right)^{\frac{W}{B}} + \frac{n_A}{g_{iA}} \frac{\rho_i}{\rho_0} \leq P_i^{\max} + \frac{n_B}{g_{iB}}, \forall i \in \mathscr{I} \right\} (2.30)$$

Meanwhile, set $\mathscr{H}_{(\rho_0)}$ is characterized by constraint (2.27) under a given ρ_0 as follows:

$$\mathscr{H}_{(\rho_0)} = \left\{ \{\rho_i\}_{i \in \mathscr{I}} \,\middle|\, \sum_{i \in \mathscr{I}} \rho_i \geq 1 - \rho_0 \right\}. \quad (2.31)$$

Note that we change the equality in (2.27) into the inequality in $\mathscr{H}_{(\rho_0)}$, and this does not change the optimal solution because of the following Proposition 2.3. Let $\{\rho_{i,(\rho_0)}^{*,\text{sub}}\}_{i \in \mathscr{I}}$ denote an optimal solution of Problem (SINR-M-SubP).

Proposition 2.3 *If Problem (SINR-M-SubP) is feasible under a given value of ρ_0, then $\sum_{i \in \mathscr{I}} \rho_{i,(\rho_0)}^{*,\text{sub}} = 1 - \rho_0$ always holds, i.e., the optimal solution of Problem (SINR-M-SubP) is consistent with (2.27).*

Top-problem to further optimize $\rho_0 \in (0, 1]$ **for minimizing the output of the subproblem** $F_{\text{sub}}(\rho_0)$ By solving the subproblem and obtaining $F_{\text{sub}}(\rho_0)$ as a function of a given ρ_0, we next solve the top-problem to compute the best ρ_0^* that can minimize the objective function:

(SINR-M-TopP): Minimize $F_{\text{sub}}(\rho_0)$, Subject to: $0 < \rho_0 \leq 1$, Variable: ρ_0.

The key advantage of the above decomposition is that we can exploit the monotonicity of Problem (SINR-M-SubP) to solve it efficiently. In the next subsection, by exploiting the above layered-structure, we design an efficient algorithm to Problem (SINR-M-P).

2.4.2 Algorithm for Optimal Offloading Solution

We first propose an efficient algorithm to solve (SINR-M-SubP) under a given ρ_0. Since the key to solve Problem (SINR-M-SubP) is the monotonic optimization theory [25, 26], we briefly introduce the monotonic optimization in this subsection. We first provide the following two important definitions.

Definition 2.1 (*Normal Set*) A set $\mathscr{G} \subset \mathscr{R}_+^n$ is normal, if for any two points x and $x' \in \mathscr{R}_+^n$ with $x' \leq x^1$ and $x \in \mathscr{G}$, we always have $x' \in \mathscr{G}$.

Definition 2.2 (*Reverse Normal Set*) A set $\mathscr{H} \subset \mathscr{R}_+^n$ is a reversed normal set, if for two points x and $x' \in \mathscr{R}_+^n$ with $x' \geq x$ and $x \in \mathscr{H}$, we always have $x' \in \mathscr{H}$.

Based on the above definitions, a canonic form of the monotonic optimization problem is as follows.

$$\max_{x} f(x), \text{ subject to: } x \in \mathscr{G} \cap \mathscr{H}, \tag{2.32}$$

where $f(x) : \mathscr{R}_+^n \to \mathscr{R}$ is an increasing function.[2] Set $\mathscr{G} \subset [0, b]$ is a normal set with nonempty interior, and set $\mathscr{H}$ is a reverse normal set on $[0, b]$, with vector b representing a given point in $\mathscr{R}_+^n$.

As shown in (2.32), the monotonic optimization problems refer to a special category of mathematical programming problems that aim to maximize a monotonic objective function subject to a feasible region constructed by the intersection of a normal set and a reversed normal set [25, 26]. The key advantage of the monotonicity is that it enables us to solve the monotonic optimization problem in a very efficient manner. Specifically, using the monotonicity of the constraints, one can consecutively construct a group of poly-blocks to approximate the feasible region closer and closer. Furthermore, thanks to the monotonicity of the objective function, the optimal solution is guaranteed to happen at one of the vertices of the constructed poly-blocks, as long as this vertex falls within (or extremely close enough to) the feasible region. Such a rationale yields an algorithm, referred as the poly-block outer-approximation algorithm, that can efficiently search for the globally optimal solution of a monotonic optimization problem.

Based on the above Definitions 2.1 and 2.2, we can show the following result.

---

[1] We say that two points x and $x' \in \mathscr{R}_+^n$ satisfy $x' \leq x$, if $x_k' \leq x_k$ holds for each element-index k of the two vectors.

[2] Given two different x and x' with $x_k \geq x_k'$, $\forall k$ and $x_j > x_j'$ for at least one index j, there always exists $f(x) < f(x')$.

Proposition 2.4 *Under a given ρ_0, Problem (SINR-M-SubP) is a monotonic optimization problem.*

Based on Proposition 2.4, we propose an algorithm, named as Algorithm (Cen-Sub), to solve Problem (SINR-M-SubP). Algorithm (Cen-Sub) is based on the rationale of poly-block outer-approximation method. Since Problem (SINR-M-SubP) aims at minimizing an increasing function, we design Algorithm (Cen-Sub) to construct a series of the poly-blocks that approximate the lower boundary of the feasible region as close as possible. Such a design is new, since the prior applications of monotonic optimization often focused on maximizing an increasing function [27].

Algorithm (Cen-Sub): to solve Problem (SINR-M-SubP) under a given $\rho_0 \in (0, 1]$

1: **Initialization:** Set the current best solution $CBS = \emptyset$, and the current best value $CBV = \infty$. Set index $k = 1$ and ϵ as a small positive number. Set the flag for stopping as $f_{\text{stop}} = 0$. Initialize set $\mathscr{T}_1 = \{\mathbf{0}\}$. We use $V(\{\rho_i\}_{i \in \mathscr{I}}) = \sum_{i \in \mathscr{I}} (\pi_B - \pi_A) W \log_2 \left(\rho_0 + \sum_{j \neq i, j \in \mathscr{I}} \rho_j \right)$ for easy presentation.

2: **while** $f_{\text{stop}} = 0$ **do**

3:     Select vertex $\mathbf{z}^k \in \arg\min \left\{ V(\{\rho_i\}_{i \in \mathscr{I}}) | \{\rho_i\}_{i \in \mathscr{I}} \in \mathscr{T}_k \right\}$.

4:     Construct a line between $\mathbf{z}^k$ and point $\mathbf{o}$ whose element $o_i = \min \left\{ 1 - \frac{1}{2^{\frac{R_i^{\text{req}}}{W}}}, \rho_0 \frac{P_{iA}^{\max} g_{iA}}{n_A}, 1 - \rho_0 \right\}, \forall i \in \mathscr{I}$.

5:     Find the intersection point $\mathbf{x}^k$ between the above constructed line and the lower boundary given in $\mathscr{H}_{(\rho_0)}$ by bisection search.

6:     **if** $V(\mathbf{x}^k) < CBV$ **then**

7:         Update $CBV = V(\mathbf{x}^k)$ and set CBS=$\mathbf{x}^k$.

8:     **end if**

9:     **if** $\| \mathbf{x}^k - \mathbf{z}^k \| < \epsilon$ **then**

10:       Set $f_{\text{stop}} = 1$.

11:     **end if**

12:     Update the set of vertexes as $\mathscr{T}_{k+1} = (\mathscr{T}_k \backslash \{\mathbf{z}^k\}) \cup \left\{ \mathbf{z}^k + (x_i^k - z_i^k)\mathbf{e}_i, i \in \mathscr{I} \right\}$.

13:     Remove all vertexes $\mathbf{z} \in \mathscr{T}_{k+1} \backslash \mathscr{G}_{(\rho_0)}$.

14:     **if** $\mathscr{T}_{k+1}$ is empty **then**

15:       Set $f_{\text{stop}} = 1$.

16:     **end if**

17:     Set $k = k + 1$.

18: **end while**

19: **Output:** Set $\{\rho_{i,(\rho_0)}^{*,\text{sub}}\}_{i \in \mathscr{I}}$ is equal to CBS, and $F_{\text{sub}}(\rho_0) = CBV$.

The key component of Algorithm (Cen-Sub) is the While-Loop (Lines 2–18), whose purpose is to iteratively construct the poly-blocks that approximate the lower boundary of $\mathscr{G}_{(\rho_0)} \cap \mathscr{H}_{(\rho_0)}$ with an increasing precision. Figure 2.3 provides a sketch to illustrate the procedures. Algorithm (Cen-Sub) terminates if $\mathbf{z}^k$ and $\mathbf{x}^k$ are close enough (Lines 9–11), or if we cannot expect to find a better solution (Lines 14–16).

We next design Algorithm (Cen) to solve Problem (SINR-M-TopP), by using Algorithm (Cen-Sub) as a subroutine to solve Problem (SINR-M-SubP) under given ρ_0. Specifically, we exploit the favorable property that Problem (SINR-M-TopP) is a

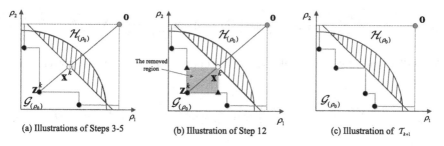

(a) Illustrations of Steps 3-5   (b) Illustration of Step 12   (c) Illustration of $\mathcal{T}_{k+1}$

Fig. 2.3 The poly-block approximation used in Algorithm (Cen-Sub). Each solid node denotes a vertex. The *shaded area* denotes the feasible region, and the *red-line* constructed by the vertices denotes the approximated lower boundary of the feasible region (color figure online)

single-variable optimization problem with a feasible region $\rho_0 \in (0, 1]$ independent on the other parameters. By using this property, we design Algorithm (Cen) to perform a line-search over $\rho_0 \in (0, 1]$ with the step-size Δ_{top} (the While-Loop on Lines 2–11). For each given value of ρ_0, we use Algorithm (Cen-Sub) to evaluate $F_{\text{sub}}(\rho_0)$ (Line 5) and update the current best solution (Line 7). We perform the feasibility-test for Problem (SINR-M-SubP) under each ρ_0 (Line 3), such that we avoid invoking Algorithm (Cen-Sub) when subproblem (SINR-M-SubP) is infeasible and hence save the computational time. In particular, we have the following property regarding Algorithm (Cen).

Proposition 2.5 *Algorithm (Cen) can yield the asymptotically optimal solution (i.e., ρ_0^* and $\{\rho_i^*\}_{i \in \mathscr{I}}$) for Problem (SINR-M-P), as Δ_{top} approaches to zero.*

Algorithm (Cen): to solve Problem (SINR-M-TopP)

1: **Initialization:** Set a small step-size Δ_{top}. Set $\rho_0 = \Delta_{\text{top}}$. Set the CBV as a very large number.
2: **while** $\rho_0 < 1$ **do**
3:    Check the feasibility of Problem (SINR-M-SubP) with Algorithm (Cen-Sub-FC) in Appendix IV.
4:    **if** Problem (SINR-M-SubP) is feasible **then**
5:        Use Algorithm (Cen-Sub) to obtain $\{\rho_{i,(\rho_0)}^{*,\text{sub}}\}_{i \in \mathscr{I}}$ and $F_{\text{sub}}(\rho_0)$.
6:        **if** $F_{\text{sub}}(\rho_0) < \text{CBV}$ **then**
7:            Set CBV $= F_{\text{sub}}(\rho_0)$. Set $\rho_0^* = \rho_0$, and $\rho_i^* = \rho_{i,(\rho_0)}^{*,\text{sub}}$, $\forall i \in \mathscr{I}$.
8:        **end if**
9:    **end if**
10:   Update $\rho_0 = \rho_0 + \Delta_{\text{top}}$.
11: **end while**
12: **Output:** ρ_0^* and $\{\rho_i^*\}_{i \in \mathscr{I}}$.

Algorithm (Cen) outputs the optimal solution $(\rho_0^*, \{\rho_i^*\}_{i \in \mathscr{I}})$ for Problem (SINR-M-P). Using $(\rho_0^*, \{\rho_i^*\}_{i \in \mathscr{I}})$, we can derive the optimal solution of Problem (CMP) as follows. Specifically, using Proposition 2.1 and Eq. (2.1), we can derive MU i's optimal transmit-power and transmission rate to the AP as

$$p_{iA}^* = \frac{n_A}{g_{iA}} \frac{\rho_i^*}{1 - \sum_{i \in \mathscr{I}} \rho_i^*} \text{ and } x_{iA}^* = W \log_2 \left(\frac{1}{1 - \rho_i^*} \right), \forall i \in \mathscr{I}.$$

With (2.8) and (2.9), we can derive MU i's optimal rate and transmit-power to the BS as

$$p_{iB}^* = \frac{n_B}{g_{iB}} \left(2^{\frac{R_i^{\text{req}}}{B}} (1 - \rho_i^*)^{\frac{W}{B}} - 1 \right) \text{ and } x_{iB}^* = R_i^{\text{req}} + W \log_2 \left(1 - \rho_i^* \right), \forall i \in \mathscr{I}.$$

Thus, we solve Problem (CMP) completely.

2.5   Numerical Results

We execute numerical experiments to show the performance of our proposed Algorithm (Cen) and the advantages of our proposed optimal offloading solution. The setup for the numerical experiments are as follows.

Network scenario and channel power gain: We setup a network scenario that the BS is located at the origin, and the small-cell AP is located at (350 m, 0 m). The MUs are randomly located within a circle, whose center is (320 m, 0 m) and the radius is 20 m. This means that the MUs are closer to the AP than to the BS (otherwise, there is little benefit of considering traffic offloading). We use the similar method as [28] to model the channel power gain, i.e., $g_{iA} = \frac{\varrho_{iA}}{l_{iA}^\kappa}$, where l_{iA} denotes the distance between MU i and the AP, and κ denotes the power-scaling factor for the path-loss (we set $\kappa = 4$). We further assume that ϱ_{iA} follows an exponential distribution with unit mean due to channel fading. Figure 2.4 plots two examples of the network scenarios (namely, an 8-MU scenario and a 12-MU scenario), which will be used in the following simulations. We set $n_0 = 1 \times 10^{-15}$ W/Hz.

Settings of radio resources: We set the AP's channel bandwidth shared by all MUs as $W = 20$ MHz (802.11a/b/g/n standard [29]) and the BS's channel bandwidth for each MU as $B = 5$ MHz (close to a WCDMA channel [30]). For each MU, we set its power capacities $P_i^{\max} = 0.25$ W (i.e., Power Class-3 of mobile devices), $P_{iA}^{\max} = 0.2$ W [30], and $P_i^{\max} = 0.35$ W. In addition, to account for the economic cost, we set $\pi_B = \$10/\text{GB}$ [31] and $\pi_A = \$2/\text{GB}$. As stated earlier, we use $\Delta_{\text{top}} = 0.005$, and we validate the choice of $\Delta_{\text{top}} = 0.005$ at the end of this section by comparing different Δ_{top}.

Accuracy and Computational Efficiency of the Proposed Algorithms: We first show the numerical results to validate the accuracy and computational efficiency of Algorithm (Cen) in Tables 2.2 and 2.3. We consider an 8-MU scenario with the randomly generated channel gains $\{g_{iA}\}_{i \in \mathscr{I}} = [0.1256, 2.8108, 0.2201, 0.0381,$ $0.5091, 0.2528, 1.4989, 0.6081] \times 10^{-4}$ and $\{g_{iB}\}_{i \in \mathscr{I}} = [2.5279, 0.6211, 1.2604,$ $0.5815, 2.5812, 1.1777, 2.6028, 2.3551] \times 10^{-8}$.

Table 2.2 presents the accuracy and computational efficiency of Algorithm (Cen) for the case of $W = 20$ MHz and $B = 5$ MHz. In Table 2.2, we vary each MU's R_i^{req}

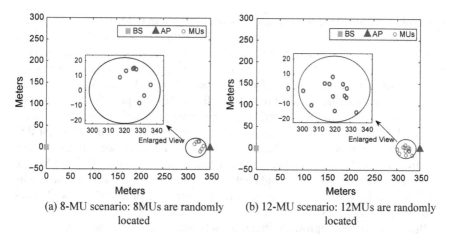

(a) 8-MU scenario: 8MUs are randomly located

(b) 12-MU scenario: 12MUs are randomly located

Fig. 2.4 Illustration of network topologies. For the sake of clear presentation, we plot the MUs' positions in an enlarged view. Subplot **a** an 8-MU scenario; Subplot **b** a 12-MU scenario

Table 2.2 Comparison between Algorithm (Cen) and LINGO ($W = 20\,\text{MHz}$, $B = 5\,\text{MHz}$): the 8-MU scenario

	$R_i^{\text{req}} =$ 2 Mbps	$R_i^{\text{req}} =$ 3 Mbps	$R_i^{\text{req}} =$ 4 Mbps	$R_i^{\text{req}} =$ 5 Mbps	$R_i^{\text{req}} =$ 6 Mbps	$R_i^{\text{req}} =$ 7 Mbps	$R_i^{\text{req}} =$ 8 Mbps
LINGO	0.032, 3649 s	0.048, 3626 s	0.073, 3668 s	0.149, 3579 s	0.225, 3614 s	0.303, 3643 s	0.381, 3629 s
Cen	0.033, 98.1 s	0.049, 120.2 s	0.074, 123.1 s	0.152, 75.7 s	0.232, 71.6 s	0.312, 68.8 s	0.393, 54.7 s

Table 2.3 Comparison between Algorithm (Cen) and LINGO ($W = 4\,\text{MHz}$, $B = 5\,\text{MHz}$): the 8-MU scenario

	$R_i^{\text{req}} =$ 1 Mbps	$R_i^{\text{req}} =$ 1.5 Mbps	$R_i^{\text{req}} =$ 2 Mbps	$R_i^{\text{req}} =$ 2.5 Mbps	$R_i^{\text{req}} =$ 3 Mbps	$R_i^{\text{req}} =$ 3.5 Mbps	$R_i^{\text{req}} =$ 4 Mbps
LINGO	0.03, 3615 s	0.069, 3609 s	0.107, 3652 s	0.146, 3587 s	0.184, 725 s	0.225, 233s	0.276, 48 s
Cen	0.031, 54.1 s	0.07, 39.7 s	0.11, 34.5 s	0.15, 28.2 s	0.189, 22.3 s	0.228, 15.7 s	0.267, 7.7 s

from 2 to 8 Mbps (as Problem (CMP) becomes infeasible when $R_i^{\text{req}} \geq 9\,\text{Mbps}$). For each cell, the first number represents the minimum total cost and the second number represents the computational time (measured in seconds and obtained by a PC with Intel Core i7-4610M CPU@3.00GHz and 8.00GB RAM). To validate Algorithm (Cen), we adopt LINGO's global-solver to solve Problem (CMP) and obtain the minimum cost.[3] Due to the non-convexity of Problem (CMP), LINGO's global-solver consumes a very long time to compute the result. The results in Table 2.2 show that Algorithm (Cen) can achieve the results very close to LINGO (with the average relative error equal to 2.89%), while consuming a significantly less computational time than LINGO. The key reason for such an advantage is that our Algorithm (Cen) exploits the layered-structure of Problem (SINR-M-P) and especially the monotonic property of Problem (SINR-M-TopP) to compute the optimal offloading solution efficiently. Table 2.3 validates the accuracy and computational efficiency of Algorithm (Cen) for the case of $W = 4\,\text{MHz}$, $B = 5\,\text{MHz}$. In Table 2.3, we vary each MU's R_i^{req} from 1 to 4 Mbps (as Problem (CMP) becomes infeasible when $R_i^{\text{req}} \geq 5\,\text{Mbps}$). Again, the results in Table 2.3 validate that Algorithm (Cen) can achieve the results very close to LINGO (with the average relative error equal to 0.67%), while consuming significantly less computational time than LINGO.

Illustration of the Optimal Offloading Solution: Second, we illustrate the MUs' optimal offloading solution by changing their traffic demands in Fig. 2.5. For the sake of clear presentation, we use a 4-MU scenario, in which the random channel power gains are $\{g_{iA}\}_{i \in \mathscr{I}} = [1.2709, 0.6407, 0.7771, 0.8638] \times 10^{-5}$ and $\{g_{iB}\}_{i \in \mathscr{I}} = [3.3164, 2.8765, 1.4029, 2.7934] \times 10^{-8}$. We vary each MU's R_i^{req} from 1 to 14 Mbps (as Problem (CMP) becomes infeasible when $R_i^{\text{req}} \geq 15\,\text{Mbps}$). In Fig. 2.5, the top-subplot shows the traffic to the AP and BS. The middle-subplot shows the total transmit-powers to the AP and BS, and the bottom-subplot shows the offloading ratio (i.e., the total traffic delivered to the AP over the total demand) and the total cost. We explain the results as follows.

- **Case of light traffic demands**: As shown in the top-subplot, when the MUs' traffic demands are low (i.e., $R_i^{\text{req}} \leq 8\,\text{Mbps}$), all MUs' demands are offloaded to the AP. Accordingly, the bottom-subplot shows the offloading-ratio equal to 1. In this situation, the MUs' transmit-powers to the AP are usually very small, as shown in the enlarged view in the middle-subplot.

- **Case of heavy traffic demands**: However, when the MUs' traffic demands further increase (i.e., $R_i^{\text{req}} \geq 9\,\text{Mbps}$), the mutual interference among the MUs becomes significant. As a result, the MUs' transmit-power constraints can no longer afford offloading all demands to the AP. As a result, some MU starts to deliver its traffic to the BS (as shown in the top-subplot), and the offloading ratio starts to decrease (as shown in the bottom-subplot). Correspondingly, the MU needs to allocate their transmit-powers to the AP and BS for accommodating its traffic scheduling, and the MUs' transmit-powers to the BS increases (as shown in the middle-subplot).

To gain deeper understanding of the results in Fig. 2.5, we plot each individual MU's traffic scheduling in Fig. 2.6a and the transmit-powers in Fig. 2.6b. The two

---

[3]LINGO is a widely used commercial optimization software to solve complicated optimization problems [3]. LINGO provides an integrated packages that can solve linear, convex, non-convex, second-order cone, and integer optimization models and etc. Since subproblem (TPA) is a non-convex optimization problem, we use the LINGO's global-solver to directly compute the optimal solution as a benchmark. The downside of using the global-solver is that it consumes a long computational time.

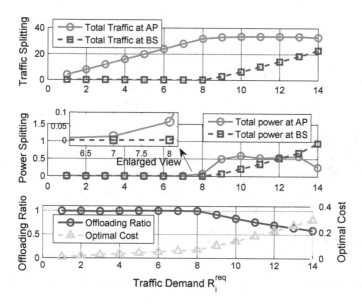

Fig. 2.5 Illustration of optimal offloading solution under different traffic demands (4-MU scenario). *Top-subplot* Traffic to the AP and BS. *Middle-subplot* Transmit-power to the AP and BS. *Bottom-subplot* Offloading ratio and the total cost

figures show that the MUs' traffic offloading decisions and the transmit-power allocations are strongly correlated due to the mutual interference at the AP. In particular, we observe that as the MUs' traffic demands increase (i.e., $R_i^{req} \geq 8$ Mbps), MU 1 (whose channel gain g_{1B} is the largest) first starts to redirect its traffic to the BS, and then MU 2 (whose g_{2B} is the second largest) follows. Such a result is consistent with the intuition, because a larger channel gain to the BS requires a smaller transmit-power to achieve the same data rate to the BS. The middle and bottom subplots of Fig. 2.6b show that each MU's transmit-power limit to the BS (i.e., $P_{iB}^{max} = 0.25$ W) and total power budget (i.e., $P_i^{max} = 0.25$ W) eventually become tight when the MUs' traffic demands increase. That is why Problem (CMP) becomes infeasible when $R_i^{req} = 15$ Mbps.

Performance Advantage of the Optimized Traffic Offloading: We show the performance advantage of the optimized traffic offloading scheme (i.e., the optimal solution of Problem (CMP)) in Fig. 2.7. We compare the performance of the optimized traffic offloading scheme with two other schemes, namely, the zero-offloading scheme and the fixed-offloading scheme. In the zero-offloading scheme, no MU's traffic is offloaded to the AP. While in the fixed-offloading scheme, each MU offloads 50% of its demand to the AP. We use the 4-MU scenario used before to execute the performance comparison. Figure 2.7 shows that the total cost can be greatly reduced by using the optimized traffic offloading scheme. In both subplots, the optimized traffic offloading can save more than 79% of the total cost compared with the zero-

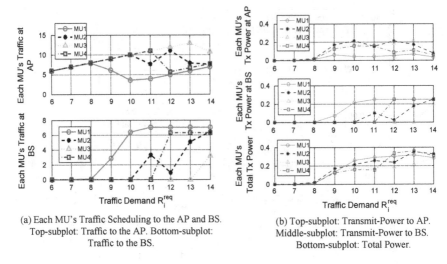

(a) Each MU's Traffic Scheduling to the AP and BS.
Top-subplot: Traffic to the AP. Bottom-subplot:
Traffic to the BS.

(b) Top-subplot: Transmit-Power to AP.
Middle-subplot: Transmit-Power to BS.
Bottom-subplot: Total Power.

Fig. 2.6 Illustration of individual MU's traffic scheduling and transmit-power allocations

offloading scheme, and more than 65% of the total cost compared with the fixed-offloading scheme.

Moreover, Fig. 2.7 validates that the optimized offloading scheme can increase the network capacity in terms of accommodating more MUs' traffic demands. Specifically, the top-subplot (for the case of $W = 20$ MHz and $B = 5$ MHz) shows that the zero-offloading cannot support a demand more than 3 Mbps for each MU, and the fixed-offloading cannot support a demand more than 7 Mbps for each MU. In comparison, the optimized traffic offloading scheme can accommodate each MU's demand up to $R_i^{req} = 14$ Mbps. Similar, the bottom-subplot (for the case of $W = 30$ MHz and $B = 10$ MHz) shows that the zero-offloading cannot support a demand more than 4 Mbps for each MU, and the fixed-offloading cannot support a demand more than 8 Mbps for each MU. In comparison, the optimized traffic offloading can accommodate each MU's demand up to $R_i^{req} = 19$ Mbps.

Performance of the Optimized Traffic Offloading Scheme under Different MUs' Locations: We evaluate the performance of the optimized offloading scheme under different locations of the MUs in Fig. 2.8. To this end, we vary the center of the circle within which the MUs are randomly located according to (170 m, 0 m), (220 m, 0 m), (270 m, 0 m), and (320 m, 0 m). This corresponds to that the MUs are gradually moving away from the BS and closer to the AP. For each location, we independently and randomly generate 100 different sets of the MUs' locations. Figure 2.8 shows the corresponding average results, in which we plot the results for three cases, i.e., each MU's demand $R_i^{req} = 4$, 8, and 12 Mbps. Subplot 2.8a shows the average total traffic offloaded to the AP versus different locations. It shows that when the MUs are closer to the AP, more traffic demands are offloaded to the AP, which is attributed to the stronger channel gains between the MUs and the AP.

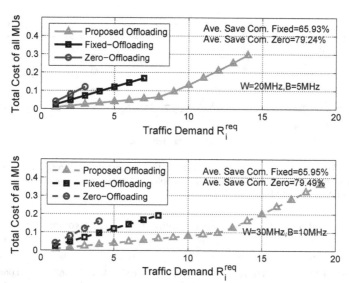

Fig. 2.7 Performance advantage of the optimized traffic offloading scheme compared with the zero-offloading scheme and the fixed-offloading scheme. *Top-subplot* case of $W = 20\,$MHz and $B = 5\,$MHz. *Bottom-subplot* case of $W = 30\,$MHz and $B = 10\,$MHz

Subplot 2.8b shows that less traffic are delivered through the BS as MUs move closer to the AP. Finally, Subplot 2.8c shows the total cost decreases when the MUs are closer to the AP due to effective offloading. An interesting observation in Subplot 2.8a is that the difference of offloaded traffic is marginal when the traffic demand increases from $R_i^{req} = 8\,$Mbps (the line marked with circles) to $R_i^{req} = 12\,$Mbps (the line marked with triangles). However, the difference is very significant when the demand increases from $R_i^{req} = 4\,$Mbps (the line marked with squares) to $R_i^{req} = 8\,$Mbps. The result is consistent with the intuition that offloading traffic to AP becomes less attractive due to the heavy interference at the AP when the MUs' traffic demands are high. In fact, as shown in Subplot 2.8b, most traffic demands are delivered to the BS when each MU's demand increases from 8 to 12 Mbps.

Impact of Stepsize Δ_{top} used in Algorithm (Cen): Finally, we evaluate the impact of Δ_{top} used in Algorithm (Cen) in Table 2.4 and validate our choice of $\Delta_{top} = 0.005$ in Algorithm (Cen) to execute the line-search. We use the 4-MU scenario and vary $\Delta_{top} = 0.001, 0.0025, 0.005,$ and 0.01 used in Algorithm (Cen). We show the obtained total cost (the first number in each cell) and the computational time (the second number in each cell) versus different traffic demands. The results in Table 2.4 show that using an Δ_{top} smaller than 0.005 yields a very limited improvement on the performance but incurs a significant increase in the computational time. Specifically, the results show that using $\Delta_{top} = 0.005$ in Algorithm (Cen) yields a small average loss of 0.385% compared to using $\Delta_{top} = 0.001$, while it can reduce 78.76% of the computational time. Similarly, compared to $\Delta_{top} = 0.0025$, using $\Delta_{top} = 0.005$

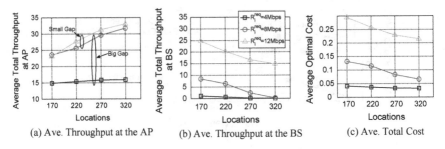

(a) Ave. Throughput at the AP   (b) Ave. Throughput at the BS   (c) Ave. Total Cost

Fig. 2.8 Performance of the proposed traffic offloading scheme under different locations of the MUs. Subplot **a** Average throughput at the AP; Subplot **b** Average throughput at the BS; Subplot **c** Average total cost of all MUs

Table 2.4 Optimal cost and computational time with different Δ_{top} (Algorithm (Cen))

	$R_i^{req} =$ 3 Mbps	$R_i^{req} =$ 4 Mbps	$R_i^{req} =$ 5 Mbps	$R_i^{req} =$ 6 Mbps	$R_i^{req} =$ 7 Mbps	$R_i^{req} =$ 8 Mbps
$\Delta_{top} = 0.01$	0.0253, 34.2 s	0.0341, 36.1 s	0.0418, 39.1 s	0.0511, 40 s	0.0565, 37.4 s	0.0666, 37.3 s
$\Delta_{top} = 0.005$	0.0253, 64.3 s	0.0327, 69.1 s	0.0404, 74.9 s	0.0497, 78.7 s	0.0565, 73.1 s	0.0651, 72.3 s
$\Delta_{top} = 0.0025$	0.0246, 106.1 s	0.0327, 112.1	0.0404, 122.4 s	0.0490, 173.9 s	0.0565, 157.9 s	0.0643, 156.4 s
$\Delta_{top} = 0.001$	0.0245, 324.9 s	0.0325, 341.1 s	0.0404, 370.5 s	0.0486, 393.5 s	0.0565, 365.4 s	0.0645, 358.9 s
	$R_i^{req} =$ 9 Mbps	$R_i^{req} =$ 10 Mbps	$R_i^{req} =$ 11 Mbps	$R_i^{req} =$ 12 Mbps	$R_i^{req} =$ 13 Mbps	$R_i^{req} =$ 14 Mbps
$\Delta_{top} = 0.01$	0.0965, 36.6 s	0.1328, 32.7 s	0.1744, 31.4 s	0.2158, 29.9 s	0.2565, 30.1 s	0.2967, 30.1 s
$\Delta_{top} = 0.005$	0.0950, 69.9 s	0.1315, 62.9 s	0.1728, 59.8 s	0.2142, 57.6 s	0.2547, 57.4 s	0.2967, 56.7 s
$\Delta_{top} = 0.0025$	0.0950, 148.9 s	0.1315, 135.5 s	0.1728, 125.5 s	0.2142, 121.9 s	0.2547, 123.9 s	0.2959, 122.7 s
$\Delta_{top} = 0.001$	0.0950, 348.1 s	0.1315, 317.2 s	0.1722, 300.3 s	0.2133, 292.1 s	0.2547, 288.6 s	0.2957, 286.9 s

yields an average loss equal to 0.288%, while reducing 46.26% of the computational time.

2.6   Extension for Bandwidth Allocation

Due to the importance of saving the spectrum usage from the cellular operators' perspective, we further extend the proposed optimal MUs' traffic offloading in this

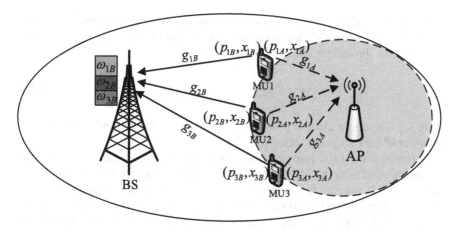

Fig. 2.9 Illustration of the system model comprised of one BS, one AP, and a group of three MUs. The BS can optimize its bandwidth allocation to the MUs according to the MUs' traffic scheduling

chapter to incorporate the BS's bandwidth allocation. Such an extension is based on the following considerations. The BS prefers to saving the licensed bandwidth usage and allocating small bandwidths to the MUs. Due to limited channel bandwidths to the macro BS, the MUs tend to offload more traffic to the small-cell AP. However, aggressively offloading traffic to the AP leads to severe co-channel interferences among the MUs, which adversely consume the MUs excessive transmit-powers. Therefore, we need to take into account a system-wise cost that accounts for both the BS's bandwidth usage as well as all MUs' mobile data cost. Specifically, we use ω_{iB} to denote the BS's bandwidth allocation to MU i. To study this extension, we consider a network scenario as shown in Fig. 2.9. Compared with Fig. 2.1, the key difference in Fig. 2.9 is that the BS now can optimize its bandwidth allocations $\{\omega_{iB}\}_{i \in \mathcal{I}}$ according to the MUs' traffic scheduling.

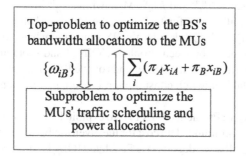

Fig. 2.10 Illustration of the layered approach to solve Problem (JOINT). Subproblem: to jointly optimize the MUs' traffic scheduling and power allocations for the given BS's bandwidth allocations to the MUs. Top-problem: to optimize the BS's bandwidth allocations based on the outcome of the subproblem

To model the BS's bandwidth allocation, we consider that the BS allocates orthogonal bands to different MUs for accommodating the uplink data as in [32, 33]. Thus, MU i's uplink throughput to the BS can be given by

$$x_{iB} = \omega_{iB} \log_2 \left(1 + \frac{p_{iB} g_{iB}}{\omega_{iB} n_0}\right), \forall i \in \mathscr{I}, \tag{2.33}$$

where g_{iB} denotes the channel power gain from MU i to the BS, and $\omega_{iB} n_0$ denotes the power of the background noise at the BS.

To optimize the radio resource usage in data offloading, we need to take into account those of both the MUs and the BS. Hence, supposing that operator owns the BS and AP, we aim at minimizing a system cost function including both the MUs' total data cost and the BS's bandwidth usage. Specifically, we formulate the following JOINT optimization problem that jointly optimizes the BS's bandwidth allocation $\{\omega_{iB}\}_{i \in \mathscr{I}}$, and the MUs' traffic scheduling $\{x_{iA}, x_{iB}\}_{i \in \mathscr{I}}$ and transmit-powers $\{p_{iA}, p_{iB}\}_{i \in \mathscr{I}}$.

(JOINT): $\min \alpha \sum_{i \in \mathscr{I}} \omega_{iB} + (1 - \alpha) \sum_{i \in \mathscr{I}} (\pi_A x_{iA} + \pi_B x_{iB})$ \hfill (2.34)

subject to: Constraints (2.1) and (2.33),

$$x_{iB} + x_{iA} \geq R_i^{\text{req}}, \forall i \in \mathscr{I}, \tag{2.35}$$

$$0 \leq p_{iA} \leq P_{iA}^{\max}, \forall i \in \mathscr{I}, \tag{2.36}$$

$$0 \leq p_{iB} \leq P_{iB}^{\max}, \forall i \in \mathscr{I}, \tag{2.37}$$

$$\omega_B^{\min} \leq \omega_{iB} \leq \omega_B^{\max}, \forall i \in \mathscr{I}, \tag{2.38}$$

variables: $\omega_{iB}, \forall i \in \mathscr{I}$, $(x_{iA}, x_{iB}), \forall i \in \mathscr{I}$, and $(p_{iA}, p_{iB}), \forall i \in \mathscr{I}$.

In (2.34), parameter α denotes the weight for the BS's bandwidth usage, and $(1 - \alpha)$ denotes the weight for the MUs' total mobile data cost. Constraint (2.35) means that MU i's traffic scheduling (x_{iA}, x_{iB}) meets its demand R_i^{req}. Constraints (2.36) and (2.37) ensure that MU i's transmit-powers to the AP and BS do not exceed the respective upper-bounds $P_{iA}^{\max}$ and $P_{iB}^{\max}$. Meanwhile, $\omega_B^{\min}$ in (2.38) denotes the BS's minimum bandwidth allocation for MU i (e.g., the minimum bandwidth for signaling-exchange with MU i), and $\omega_B^{\max}$ denotes the BS's maximum bandwidth allocation for MU i. The key difference between Problem (JOINT) and Problem (CMP) are two folded: (i) MU i's traffic rate to the BS depends on the BS's bandwidth allocation ω_{iB}, and (ii) the BS's total bandwidth consumption is taken into account in the objective function.

The key idea to solve Problem (JOINT) is to exploit the proposed Algorithm (Cen) to compute the MUs' minimum data cost for the given the BS's bandwidth allocations to the MUs. Specifically, we decompose Problem (JOINT) into two subproblems as shown in Fig. 2.10. The subproblem is to optimize the MUs' traffic scheduling and power allocations for minimizing the MUs' total mobile data cost, under the given BS's bandwidth allocations to the MUs. This subproblem is similar to Problem

(CMP), and Algorithm (Cen) is applicable to compute the optimal solution. Next, the top-problem is to further optimize the BS's bandwidth allocations to the MUs based on the outcome of subproblem, with the objective of minimizing the system-wise cost.

2.7 Summary

In this chapter, we have proposed an optimal resource allocation for the small-cell-based MUs' uplink traffic offloading with dual connectivity. Offloading traffic to the small cell can help reduce the MUs' mobile data cost, but at the expense of suffering increased interferences from other MUs accessing to the same small cell. Thus, we aim at minimizing the MUs' total mobile data cost while controlling the co-channel interference. To study the problem, we formulated an optimization problem that jointly determines the MUs' traffic scheduling and power allocations to the macro-cell BS and small-cell AP, and further proposed an efficient algorithm to solve the joint optimization problem and compute the optimal offloading solution. Numerical results have showed that the accuracy and computational efficiency of the proposed algorithm. Moreover, numerical results show that the optimized traffic offloading scheme can significantly reduce the MUs' total mobile data cost and enhance the offloading capacity, in comparison with some heuristic offloading schemes. Finally, based on the study in this chapter, we have further showed the extension that incorporates the BS's bandwidth usage in the optimal resource allocation for the small-cell-based traffic offloading with DC.

References

1. N. S. Networks, "LTE Release 12 and Beyond," http://networks.nokia.com/system/files/document/nokia_lte_a_evolution_white_paper.pdf.
2. N. Ali, A. Taha, and H. Hassanein, "Quality of Service in 3GPP R12 LTE-Advanced," *IEEE Communications Magazine*, vol. 51, no. 8, pp. 103–109, 2013.
3. L. Schrage, *Optimization Modeling with LINGO.* Lindo System, 1999.
4. Y. Wu, Y. He, L. Qian, J. Huang, and X. Shen, "Joint scheduling and power allocations for traffic offloading via dual-connectivity," http://arxiv.org/abs/1509.09241, Sep. 2015.
5. S. Jha, K. Sivanesan, R. Vannithamby, and A. Koc, "Dual connectivity in LTE small cell networks," in *Proc. of IEEE GLOBECOM*, Austin, TX, Dec. 2014.
6. A. Mukherjee, "Macro-small cell grouping in dual connectivity LTE-B networks with non-ideal backhaul," in *Proc. of IEEE ICC*, Sydney, Australia, Jun. 2014.
7. J. Liu, J. Liu, and H. Sun, "An enhanced power control scheme for dual connectivity," in *Proc. of IEEE VTC-Fall*, Vancouver, Canada, Sep. 2014.
8. A. Mukherjee, "Optimal flow bifurcation in networks with dual base station connectivity and non-ideal backhaul," in *Proc. of Asilomar Conference on Signals, Systems and Computers*, Pacific Grove, CA, Nov. 2014.
9. H. Wang, C. Rosa, and K. Pedersen, "Dual connectivity for LTE-Advanced heterogeneous networks," *Wireless Networks*, vol. 22, no. 4, pp. 1315–1328, 2016.

10. F. Zhang, W. Zhang, and Q. Ling, "Non-cooperative game for capacity offload," *IEEE Transactions on Wireless Communications*, vol. 11, no. 4, pp. 1565–1575, 2012.
11. Z. Wang and V. Wong, "A novel D2D data offloading scheme for LTE networks," in *Proc. of IEEE ICC*, London, UK, May 2015.
12. Q. Ye, B. Rong, Y. Chen, M. Al-Shalash, C. Caramanis, and J. Andrews, "User association for load balancing in heterogeneous cellular networks," *IEEE Transactions on Wireless Communications*, vol. 12, no. 6, pp. 2706–2716, 2013.
13. C. Ho, D. Yuan, and S. Sun, "Data offloading in load coupled networks: A utility maximization framework," *IEEE Transactions on Wireless Communications*, vol. 13, no. 4, pp. 1912–1931, 2014.
14. X. Chen, J. Wu, Y. Cai, H. Zhang, and T. Chan, "Energy-efficiency oriented traffic offloading in wireless networks: A brief survey and a learning approach for heterogeneous cellular networks," *IEEE Journal on Selected Areas in Communications*, vol. 33, no. 4, pp. 627–640, 2015.
15. G. Iosifidis, L. Gao, J. Huang, and L. Tassiulas, "Double-auction mechanism for mobile data-offloading markets," *IEEE/ACM Transactions on Networking*, vol. 23, no. 5, pp. 1634–1647, 2015.
16. Y. Yang, T. Quek, and L. Duan, "Backhaul-constrained small cell networks: Refunding and QoS provisioning," *IEEE Transactions on Wireless Communications*, vol. 13, no. 9, pp. 5148–5161, 2014.
17. X. Kang, Y. Chia, S. Sun, and H. Chong, "Mobile data offloading through a thrid-party WiFi access point: An operator's perspective," *IEEE Transactions on Wireless Communications*, vol. 13, no. 10, pp. 5340–5351, 2014.
18. Y. Wu, K. Guo, L. P. Qian, J. Wang, and W. Lu, "Joint access-selection and power allocation for mobile data offloading in cellular networks," in *Proc. of IWCMC*, Cyprus, Sep. 2016.
19. N. Networks, "Future work: Optimizing spectrum utilisation towards 2020," http://networks.nokia.com/file/30301/optimising-spectrum-utilisation-towards-2020.
20. S. Guruacharya, D. Niyato, D. I. Kim, and E. Hossain, "Hierarchical competition for downlink power allocation in OFDMA femtocell networks," *IEEE Transacations on Wireless Communications*, vol. 12, no. 4, pp. 1543–1553, 2013.
21. M. Cheung, R. Southwell, and J. Huang, "Congestion-aware network selection and data offloading," in *Proc. of IEEE CISS*, Princeton, NJ, March 2014.
22. L. Gao, G. Iosifidis, J. Huang, and L. Tassiulas, "Economics of mobile data offloading," in *Proc. of IEEE INFOCOM Workshops*, Turin, April 2013.
23. J. Huang, R. Berry, and M. L. Honig, "Distributed interference compensation for wireless networks," *IEEE Journal on Selected Areas in Communications*, vol. 24, no. 5, pp. 1074–1084, 2006.
24. M. Chiang, C. Tan, D. Palomar, D. Neill, and D. Julian, "Power control by geometric programming," *IEEE Transactions on Wireless Communications*, vol. 6, no. 7, pp. 2640–2651, 2007.
25. Y. Zhang, L. Qian, and J. Huang, *Monotonic Optimization in Communication and Networking Systems*. Now Publisher, 2013.
26. H. Tuy, "Monotonic optimization: Problems and solution approaches," *SIAM Journal of Optimization*, vol. 11, no. 2, pp. 464–494, 2000.
27. L. Qian, Y. J. Zhang, and J. Huang, "MAPEL: Achieving global optimality for a non-convex wireless power control problem," *IEEE Transactions on Wireless Communications*, vol. 8, no. 3, pp. 1553–1563, 2009.
28. R. Zhang, "Optimal dynamic resource allocation for multi-antenna broadcasting with heterogeneous delay-constrained traffic," *IEEE Journal of Selected Topics in Signal Processing*, vol. 2, no. 2, pp. 243–255, 2008.
29. O. Bejarano and E. Knightly, "IEEE 802.11ac: From channelization to multi-user MIMO," *IEEE Communications Magazine*, vol. 51, no. 10, pp. 84–90, 2013.
30. N. Instruments, "Introduction to UMTS device testing transmitter and receiver measurements for WCDMA devices," http://download.ni.com/evaluation/rf/Introduction_to_UMTS_Device_Testing.pdf.

31. S. Ha, C. Wong, S. Sen, and M. Chiang, "Pricing by timing: Innovating broadband data plans," in *Proc. of SPIE OPTO Broadband Access Communication Technologies VI Conference*, California, USA, Jan. 2012.

32. J. Huang, V. Subramanian, R. Agrawal, and R. Berry, "Joint scheduling and resource allocation in uplink OFDM systems for broadband wireless access networks," *IEEE Journal on Selected Areas in Communications*, vol. 27, no. 2, pp. 226–234, Feb. 2009.

33. G. Yu, Y. Jiang, L. Xu, and G. Li, "Multi-objective energy-efficient resource allocation for multi-RAT heterogenous networks," *IEEE Journal on Selected Areas in Communications*, vol. 33, no. 10, pp. 2118–2127, Oct. 2015.

Chapter 3
Resource Allocation for D2D-Assisted Traffic Offloading

With the explosive growth of mobile applications, people are relying more heavily on mobile devices for sharing contents and watching video streaming. Recent report from Cisco has predicted that mobile video will grow at a Compound Annual Growth Rate (CAGR) of 62% between 2015 and 2020, and even three-fourths of mobile traffic will be video streaming by 2020 [1]. Therefore, it is of practical importance for cellular operators to accommodate the mobile traffic (for content distribution and video streaming) in a cost-efficient manner. Offloading traffic based on mobile users (MUs') device-to-device (D2D) cooperation provides a promising solution to achieve this goal. Given a group of MUs in close proximity and interested in downloading a common video clip, the cellular Base Station (BS) can send part of data to some selected MUs, who then cooperatively offload the received data to other neighboring MUs via the D2D-links. Thanks to short distance among the MUs, such a D2D-assisted traffic offloading can significantly improve radio resource utilization for content distribution, e.g., lowering the energy consumption and reducing the BS's bandwidth usage [2–6].

In this chapter, we study the optimal resource allocation for the MUs' D2D-assisted downlink traffic offloading. Specifically, we focus on a scenario in which a group of neighboring MUs aim at downloading a common piece content (e.g., video stream) from the BS. In the considered offloading process, the BS first actively sends data to some selected MUs who then offload the received data to other MUs via D2D-links, such that all MUs can receive the entire content. To achieve the benefit of D2D-assisted traffic offloading, we need to properly design the resource allocations for the following two reasons. First, due to each MU's limited transmit power and energy budget, content transmission control will influence the MUs' cooperations, namely, which MUs can be selected for offloading and for how long. Second, the content transmission control directly influences the BS's cellular-link usage. In other words, there exists a complicated coupling among the content transmission control, the MUs' offloading strategies, and the consequent radio resource consumption. To investigate this problem, we formulate a joint optimization of the content transmission and the MUs' offloading durations to minimize the system cost function that accounts

© The Author(s) 2017
Y. Wu et al., *Radio Resource Management for Mobile Traffic Offloading in Heterogeneous Cellular Networks*, SpringerBriefs in Electrical and Computer Engineering, DOI 10.1007/978-3-319-51037-8_3

for the total energy consumption as well as the BS's cellular-link usage. The detailed contributions in this chapter are summarized as follows.

- **Optimization Formulation**: We formulate an optimization problem that jointly controls the transmission rate of the content and each MU's offloading duration, with the objective of minimizing the system-wise cost while satisfying each MU's QoS requirement. The system-wise cost accounts for the BS's and all MUs' energy consumption cost and the BS's cellular-link usage cost.
- **Algorithm Design**: We characterize the optimal solution of the joint optimization problem and propose an efficient algorithm to compute the solution. The joint optimization problem is difficult to solve, since the transmission rate of the content influences each MU's offloading duration in a complicated manner, which yields a difficult nonconvex optimization problem. To tackle with this difficulty, we identify the decomposition property of the joint optimization, based on which we characterize different possible cases when achieving the optimum. We then analytically derive the corresponding optimal solution for each of these cases, and finally propose an efficient algorithm to find the optimal solution of the original joint optimization problem.
- **Performance Evaluation**: We perform extensive numerical simulations to validate the derived analytical results and the proposed algorithm to compute the optimal offloading solution. In particular, the results show that the proposed algorithm saves more than 90% of the computational time compared to an exhaustive search method. Besides, we show through simulations that the D2D-assisted offloading with the jointly optimized content transmission rate and the MUs' offloading durations can significantly reduce the overall system cost, in comparison with some other heuristic offloading schemes.

The remainder of this chapter is organized as follows. We first review the related studies in Sect. 3.1. Then, we present the system model in Sect. 3.2, and illustrate the joint optimization problem in Sect. 3.3. We propose an efficient algorithm to compute the optimal offloading solution in Sect. 3.4, and further show the numerical results in Sect. 3.5. Finally, we discuss about the extension of the optimized D2D-assisted traffic offloading in Sect. 3.6, and summarize this chapter in Sect. 3.7. The complete results of this chapter and all the technical proofs can be found in [7, 8].

3.1   Related Studies

Related studies about resource allocation for D2D-assisted traffic offloading can be roughly categorized into the following two groups.

The first group of studies focused on the D2D-assisted offloading for real-time traffic. These results mainly investigated which MUs offload which parts of the contents, with the objective of optimizing different performances such as the energy consumption and the cellular-link usage. Al-Kanj et al. in [9] investigated how to separate MUs into different groups and select one leader of each group for offloading content, with the objective of minimizing the total power consumption. Further in [10], Al-Kanj et al. investigated the grouping and leader-selection problem for minimizing the bandwidth usage of the cellular link. Wang et al. in [11] considered vehicle-to-vehicle assisted traffic offloading, and proposed a coalition game based approach for distributing contents within a group of vehicles. An involved issue is how to incentivize the MUs to execute the D2D-assisted offloading. In [12], Gao et al. proposed a hybrid pricing reimbursing policy for motivating the MUs to play as hosts providing Internet connections. In [13], Vu et al. proposed a tit-for-tat mechanism to motivate the MUs' peer-to-peer cooperations. In [14], Niyato et al. proposed a sequential game model to analyze the cooperations between network operators and content providers for content delivery.

The second group of studies focused on the D2D-assisted offloading for delay-tolerant traffic. The main focus is to design different opportunistic offloading schemes for content distribution. In [15, 16], different schemes were proposed to migrate delay-tolerant traffic from cellular links to D2D or WiFi links. In [17], Li et al. proposed an energy-efficient opportunistic forwarding scheme for maximizing the message delivery probability. In [18], Wang et al. proposed a hybrid pull-and-push scheme for opportunistic D2D-assisted content delivery. In [19], Whitbeck et al. considered a push-based architecture for opportunistic content delivery via users' cooperations and evaluated the influence of the number of content copies. In [20], Golrezaei et al. proposed a D2D-assisted femto-caching scheme for video distribution. Moreover, a multi-hop broadcasting protocol has been proposed in [21] for message distribution in urban vehicular networks.

Our study here belongs to the first group of studies. Different from those studies in the first group that mainly modeled and analyzed the MUs' D2D-cooperations from a macro view, our study this chapter takes a micro and analytical approach for modeling and optimizing the radio resource usage for the MUs' D2D-assisted offloading. We focus on investigating the coupling effect between the content transmission control and the MUs' offloading strategies. As we described before, such a coupling effect significantly influences the resource usage and the consequent offloading performance. Moreover, taking into account the coupling effect leads to a challenging nonconvex resource allocation problem, which requires an efficient algorithm to compute the optimal offloading solution.

3.2 System Model

3.2.1 MUs' Cooperative Offloading Through D2D-Links

We consider a system model comprised of a set $\mathscr{I} = \{1, 2, ..., I\}$ of MUs who are in close proximity and interested in downloading a common piece of content from the BS (as shown in Fig. 3.1a). The size of the content is L bits, and the correspond-

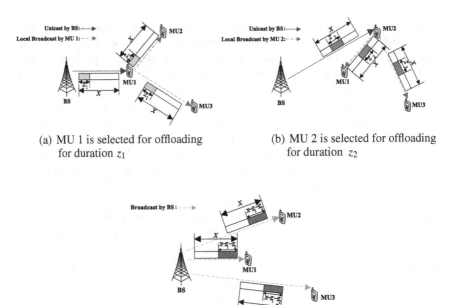

(a) MU 1 is selected for offloading   (b) MU 2 is selected for offloading
         for duration z_1 for duration z_2

(c) The BS broadcasts for duration $x - z_1 - z_2$

Fig. 3.1 Illustration of the considered cooperative scheme among three MUs, i.e., $\mathscr{I} = \{1, 2, 3\}$. The content transmission duration is x. MU 1 and MU 2 offload the data for offloading durations z_1 and z_2, respectively. The BS broadcasts to all MUs for the duration $x - z_1 - z_2$ to finish delivering the content

ing transmission rate is r bits/second. Hence, the total transmission duration of the content is $x = L/r$.

The considered MUs' D2D-assisted offloading works as follows. The BS first sends part of the data to a selected MU i at a rate r for a duration z_i. Meanwhile, MU i offloads its received data to other MUs at the same rate r. Followed by this procedure, the BS sequentially chooses different MUs for offloading data, and the selected MUs then offload their received data to the other MUs. We consider that the MUs use the same transmission rate as the BS, which corresponds to a benchmark case that each MU uses the same coding rate/scheme (as that of the BS) for offloading the content, without invoking adaptive coding scheme. Figure 3.1a illustrates the case that MU 1 offloads for duration z_1. The blue solid arrow represents the unicast transmission from the BS, and the two red dash lines represent the broadcast-transmission from MU 1 for offloading its received data. Figure 3.1b illustrates the case that MU 2 performs the offloading for duration z_2. Due to the MUs' limited transmit powers and energy capacities, the BS might need to broadcast some data to all MUs to finish the delivery of the whole content. The duration for the BS to broadcast, if needed, is $x - \sum_{i \in \mathscr{I}} z_i$. Figure 3.1c shows the case that the BS broadcasts to all MUs (i.e., the green dash lines). However, the BS's broadcast-transmission is undesirable, since it consumes a significant transmit power due to taking account of the MU with the worst channel condition from the BS.

Based on the above offloading model, we aim to jointly optimize the content transmission rate r (or equivalently its transmission duration x) and each MU i's offloading duration z_i, in order to minimize the total system cost for the content delivery. The total system cost includes three parts: (i) the energy consumption of the BS, (ii) the energy consumption of each MU, and (iii) the usage of the cellular link. The details of the modeling are presented in the next subsection.

3.2.2 Models of Resource Consumption

We model resource usages in the considered offloading model, which include the BS's energy consumption, each MU's energy consumption, and the usage of BS's cellular link. Table 3.1 lists all the important notations used.

(i) **Model of BS's Energy Consumption**: We first model the BS's energy consumption, which includes two parts, i.e., that for unicasting to MU i (when MU i is selected for offloading data), and that for broadcasting to all MUs. We describe the details as follows:

(i.1) **BS's Energy Consumption for Unicasting**: Suppose that MU i is selected by the BS for offloading with duration z_i. During z_i, the BS unicasts to MU i at the transmission rate r. We use $F_{Bi}(r)$ to denote the required transmit power by the BS to perform this unicasting (the subscript "B" stands for the BS), and such power depends on the choice of MU i (hence, the subscript i is included). Using the Shannon's channel capacity formula, $F_{Bi}(r)$ can be expressed as $F_{Bi}(r) = (2^r - 1)n/g_{Bi}$, where for the sake of clear presentation, we assume an unit bandwidth of the channel. Parameter g_{Bi} denotes the channel power gain from the BS to MU i, and parameter n denotes the power of the background noise. In addition to the transmit power, the BS also consumes a static circuit power dissipation q_B when it is transmitting data. We assume that q_B is independent of the transmit power. Hence, the BS's total energy consumption when it selects MU i for offloading with duration z_i is given by

$$E_{Bi}(x, z_i) = \left(F_{Bi}(r) + q_B\right)z_i = (2^r - 1)\frac{n}{g_{Bi}}z_i + q_B z_i. \qquad (3.1)$$

(i.2) **BS's Energy Consumption for Broadcasting**: When $x - \sum_{i \in \mathscr{I}} z_i > 0$, the BS needs to finish the content transmission by broadcasting to all MUs. We use function $F_{B0}(r) = (2^r - 1)\frac{n}{\min_{i \in \mathscr{I}}\{g_{Bi}\}}$ to denote the required transmit power of the BS for successfully broadcasting to all the MUs at the transmission rate r. The $\min_{i \in \mathscr{I}}\{g_{Bi}\}$ in the denominator is due to the fact that the broadcasting of the BS should take into account the MU with the worst channel power gain. Then, the BS's energy consumption to perform such a broadcasting is given by

Table 3.1 Important parameters and variables

$\mathscr{I}$	The set of MUs $\mathscr{I} = \{1, 2, ..., I\}$	L	The size of content
$T^{\max}$	Prefixed upper bound of transmission duration	$P_i^{\max}$	The power capacity of MU i
q_B	The circuit power consumption when the BS is transmitting	h_i	MU i's reception power consumption when receiving content
z_i	The offloading duration of MU i	$F_{Bi}(r)$	The transmit power that the BS unicasts content to MU i at bit rate r
x	The transmission duration of content	$F_{B0}(r)$	The transmit power that the BS broadcasts content to all the MUs
g_{Bi}	The channel gain of the channel form the BS to MU i	$F_i(r)$	Required transmit power of MU i for broadcasting to the other MUs
g_{ij}	The channel gain of the channel form MU i to MU j	$E_{Bi}(x, z_i)$	The total energy consumption of the BS when selecting MU i for offloading with a duration z_i
$E_i^{\text{rec}}(x)$	The energy consumption of MU i for receiving the whole content	$E_{B0}(x, \{z_i\}_{i \in \mathscr{I}})$	The energy consumption for the BS to perform broadcasting
$E_i^{\text{tot}}(x, z_i)$	The total energy consumption of MU i	$E_B^{\text{tot}}(x, \{z_i\}_{i \in \mathscr{I}})$	The BS's total energy consumption
E_i^{b}	MU i's energy consumption budget	α	Weight of the energy consumption of the BS
n	The noise power of the channel	β_i	Weight of the energy consumption of each MU i
γ	Weight of the cost for the cellular-link usage		

$$E_{B0}(x, \{z_i\}_{i \in \mathscr{I}}) = \left(F_{B0}(r) + q_B\right)\left(x - \sum_{i \in \mathscr{I}} z_i\right) = \left((2^r - 1)\frac{n}{\min_{i \in \mathscr{I}}\{g_{Bi}\}} + q_B\right)\left(x - \sum_{i \in \mathscr{I}} z_i\right).$$
$$(3.2)$$

Summarizing (3.1) and (3.2), the BS's total energy consumption is given by

$$E_B^{\text{tot}}(x, \{z_i\}_{i \in \mathscr{I}}) = E_{B0}(x, \{z_i\}_{i \in \mathscr{I}}) + \sum_{i \in \mathscr{I}} E_{Bi}(x, z_i). \qquad (3.3)$$

(ii) Model of each MU's Energy Consumption: We next model each MU's energy consumption. Each MU i's energy consumption includes two parts, i.e., that for data reception, and that for offloading the received data. The details are as follows:

(ii.1) MU's Energy Consumption for Data Reception: The main operation of the MUs is data reception. The circuit power consumption of each MU when it is receiving data can be modeled as a constant, and we denote it by h_i for MU i. There exist three possible scenarios in which MU i is receiving data, namely, (a) when the BS unicasts the data to MU i (when MU i is selected to offload data), (b) when some other MU $i' \neq i$ broadcasts to MU i (when MU i' is selected to offload data), and (c) when the BS broadcasts to all MUs. Considering these three scenarios, MU i's energy consumption for receiving the whole content is given by

$$E_i^{\text{rec}}(x) = h_i \left(\left(x - \sum_{i \in \mathscr{I}} z_i \right) + z_i + \sum_{i' \neq i, i' \in \mathscr{I}} z_{i'} \right) = h_i x. \qquad (3.4)$$

(ii.2) MU's Energy Consumption for Offloading: If selected, MU i offloads its received data to the other MUs for duration z_i. We use the following function:

$$F_i(r) = (2^r - 1) \frac{n}{\min_{i' \neq i, i' \in \mathscr{I}} \{g_{ii'}\}} \qquad (3.5)$$

to denote the required transmit power of MU i for broadcasting to the other MUs (where $g_{ii'}$ is the channel gain from MU i to a different MU i'). Thus, MU i's energy consumption for offloading its received data to all the other MUs is equal to $(F_i(r) + q_i)z_i$. In practice, the circuit power consumption of mobile user (when transmitting) is usually significantly smaller than that of the cellular base station. For example, according to [22–26], the circuit power consumption of cellular base stations when transmitting is around the order of 1-10W. In comparison, according to [9, 27, 28], the circuit power power of mobile users when transmitting is around the order of 10mW, which is 1% or less than that of the BS. Therefore, for simplicity, we do not explicitly consider q_i in each MU's energy consumption.

Summarizing the above two parts, MU i's total energy consumption is given by

$$E_i^{\text{tot}}(x, z_i) = E_i^{\text{rec}}(x) + F_i(r)z_i. \qquad (3.6)$$

3.3 Problem Formulation and Decomposable Structure

3.3.1 Problem Formulation

We formulate an optimization problem to jointly optimize the content's transmission rate r (or equivalently, its transmission duration x) and each MU i's offloading duration z_i as follows (the capital letters "JTCOO" refer to "Joint Transmission Control and Offloading Optimization"):

$$(\text{JTCOO}): \quad \min_{x, r, \{z_i\}_{i \in \mathscr{I}}} O(x, \{z_i\}_{i \in \mathscr{I}}) = \alpha E_B^{\text{tot}}(x, \{z_i\}_{i \in \mathscr{I}}) + \sum_{i \in \mathscr{I}} \beta_i E_i^{\text{tot}}(x, z_i) + \gamma x$$

subject to: $\quad x = \dfrac{L}{r},$ \hfill (3.7)

$$x \leq T^{\max}, \hfill (3.8)$$

$$\sum_{i \in \mathscr{I}} z_i \leq x, \hfill (3.9)$$

$$0 \leq z_i \leq x \mathbb{I}(F_i(r) \leq P_i^{\max}), \forall i \in \mathscr{I}, \hfill (3.10)$$

$$E_i^{\text{tot}}(x, z_i) \leq E_i^{\text{b}}, \forall i \in \mathscr{I}, \hfill (3.11)$$

variables: $\quad x, \{z_i\}_{i \in \mathscr{I}}.$

Problem (JTCOO) aims to minimize the total system cost that includes the total energy consumption of the BS and all MUs as well as the cellular-link usage. The first two terms in $O(x, \{z_i\}_{i \in \mathscr{I}})$ capture the energy consumption of the BS (weighted by α) and that of each MU i (weighted by β_i). The third term in $O(x, \{z_i\}_{i \in \mathscr{I}})$ accounts for the cost for the cellular-link usage (weighted by γ). Constraint (3.8) ensures that the transmission duration x cannot exceed a prefixed upper bound $T^{\max}$, which corresponds to the strict deadline for delivering the content. In Problem (JTCOO), the transmission duration of the content influences the MUs' offloading durations $\{z_i\}_{i \in \mathscr{I}}$ in a complicated manner. Constraint (3.9) ensures that all MUs' total offloading duration cannot exceed the transmission duration x. Constraint (3.10) means that MU i is eligible for offloading, only if its required transmit power for broadcasting $F_i(r)$ in (3.5) is below its transmit power limit $P_i^{\max}$. Here, the indicator function $\mathbb{I}(\mu) = 1$ if condition μ is satisfied, and $\mathbb{I}(\mu) = 0$ otherwise. Constraint (3.11) ensures that MU i's total energy consumption $E_i^{\text{tot}}(x, z_i)$ in (3.6) cannot exceed its energy consumption budget E_i^{b} (where the superscript "b" represents "budget"). Each MU i sets its own energy budget E_i^{b} based on its own interest in offering the offloading. A small budget E_i^{b} implies that MU i focuses on receiving the content only, and a large budget E_i^{b} implies that MU i is willing to offer the offloading. We notice that Problem (JTCOO) is always feasible, since at least the BS can send the content to all MUs via broadcasting without invoking any MU's offloading. We use x^* (which leads to $r^* = L/x^*$) and $\{z_i^*\}_{i \in \mathscr{I}}$ to denote the optimal solution of Problem (JTCOO) (i.e., the optimal offloading solution).

3.3.2 Decomposable Structure

Problem (JTCOO) is a nonconvex optimization problem, and there exists no generic algorithm that can efficiently compute the optimal offloading solution. In particular, we notice that x determines each MU's required transmit power to execute the consequent offloading, which thus influences (i) whether MU i is eligible to be selected for offloading and (ii) how long MU i can offload. This motivates us to exploit decomposable structure of Problem (JTCOO). Specifically, we can express function $O(x, \{z_i\}_{i \in \mathscr{I}})$ as follows:

$$O(x, \{z_i\}_{i \in \mathscr{I}}) = \sum_{i \in \mathscr{I}} \left(\alpha F_{Bi} \left(\frac{L}{x} \right) - \alpha F_{B0} \left(\frac{L}{x} \right) + \beta_i F_i \left(\frac{L}{x} \right) \right) z_i + \alpha \left(F_{B0} \left(\frac{L}{x} \right) + q_B \right) x$$

$$+ \sum_{i \in \mathscr{I}} \beta_i h_i x + \gamma x, \tag{3.12}$$

in which only the first term depends on $\{z_i\}_{i \in \mathscr{I}}$. Using the property, we decompose Problem (JTCOO) into two subproblems as follows:

Subproblem to optimize $\{z_i\}_{i \in \mathscr{I}}$ **under given** x: It is easy to see from (3.4), (3.6), and (3.11) that the transmission duration of the content x cannot exceed $\min_{i \in \mathscr{I}} \{ \frac{E_i^b}{h_i} \}$, otherwise some MU will violate its energy budget constraint even by just receiving. Together with (3.8), we can limit x in the interval of $\left[0, \min\{\min_{i \in \mathscr{I}} \{E_i^b/h_i\}, T^{\max}\} \right]$. Supposing that the value of x is given, we obtain the bottom-layer subproblem that optimizes each MU's offloading durations $\{z_i\}_{i \in \mathscr{I}}$ as follows:

(JTCOO-Bottom): $O_{\text{bot}}(x) = \min\limits_{\{z_i\}_{i \in \mathscr{I}}} \sum\limits_{i \in \mathscr{I}} \left(\alpha F_{Bi} \left(\frac{L}{x} \right) - \alpha F_{B0} \left(\frac{L}{x} \right) + \beta_i F_i \left(\frac{L}{x} \right) \right) z_i$

subject to: $\displaystyle\sum_{i \in \mathscr{I}} z_i \leq x,$ (3.13)

$$0 \leq z_i \leq x \mathbb{I} \left(F_i \left(\frac{L}{x} \right) \leq P_i^{\max} \right), \forall i \in \mathscr{I}, \tag{3.14}$$

$$F_i \left(\frac{L}{x} \right) z_i \leq E_i^b - h_i x, \forall i \in \mathscr{I}, \tag{3.15}$$

variables: $\{z_i\}_{i \in \mathscr{I}}.$

In Problem (JTCOO-Bottom), we have replaced r by x via using (3.7). We denote the optimal value of the bottom Problem (JTCOO-Bottom) as $O_{\text{bot}}(x)$, which depends on x. We will analytically drive $O_{\text{bot}}(x)$ in Sect. 3.4.

Top-problem to optimize x: After deriving $O_{\text{bot}}(x)$, we can continue to solve the top-layer subproblem that optimizes x as follows:

(JTCOO-Top): $\min\limits_{x} O_{\text{bot}}(x) + \alpha \left(F_{B0} \left(\frac{L}{x} \right) + q_B \right) x + \sum\limits_{i \in \mathscr{I}} \beta_i h_i x + \gamma x$

subject to: $0 \leq x \leq X^{\text{up}} = \min \left\{ \min\limits_{i \in \mathscr{I}} \{ \frac{E_i^b}{h_i} \}, T^{\max} \right\},$ (3.16)

variables: $x.$

By solving Problem (JTCOO-Bottom) and Problem (JTCOO-Top) in a way of backward induction, we can solve the original Problem (JTCOO).

3.4   Optimal Offloading Solution and Proposed Algorithm

3.4.1   Optimal Offloading Duration: Solution of Subproblem

In this section, we focus on solving Problem (JTCOO-Bottom). Under a fixed value of x, the objective function and constraints (3.13)–(3.15) of Problem (JTCOO-Bottom) are linear with respect to the decision variables $\{z_i\}_{i \in \mathscr{I}}$. Therefore, Problem (JTCOO-Bottom) is a linear programming problem. To avoid confusion, we use $\{z_i^{\text{bot}}(x)\}_{i \in \mathscr{I}}$ to denote the optimal solution of Problem (JTCOO-Bottom), which depends on the given x. To derive $\{z_i^{\text{bot}}(x)\}_{i \in \mathscr{I}}$, we first introduce parameter M_i of each MU i as follows:

$$M_i = \alpha \frac{n}{g_{Bi}} + \beta_i \frac{n}{\min_{i' \neq i, i' \in \mathscr{I}}\{g_{ii'}\}} - \alpha \frac{n}{\min_{i' \in \mathscr{I}}\{g_{Bi'}\}}, \forall i \in \mathscr{I}. \qquad (3.17)$$

Parameter M_i is an important parameter that indicates how *helpful* MU i is in offloading. For the sake of easy presentation, we make the following assumption in the rest of the chapter.

Assumption 1 (*An initial ordering of the MUs*) All MUs in $\mathscr{I}$ have been ordered according to an ascending order based on $\{M_i\}_{i \in \mathscr{I}}$, i.e.,

$$M_1 \leq M_2 \leq ... M_N < 0 \leq M_{N+1} \leq ... \leq M_I, \qquad (3.18)$$

always holds, where parameter N denotes the number of MUs whose $M_i < 0$.

Based on Assumption 1, we can derive $\{z_i^{\text{bot}}(x)\}_{i \in \mathscr{I}}$ in the following proposition (all technical proofs in this chapter can be found in [7]):

Proposition 3.1 *The optimal solution of Problem (JTCOO-Bottom) under a given value of x is as follows. For each MU i with $1 \leq i \leq N$, its unique optimal offloading duration is*

$$z_i^{\text{bot}}(x) = \min \left\{ \eta_i \mathbb{I} \left(F_i\left(\frac{L}{x}\right) \leq P_i^{\max} \right), \frac{E_i^b - h_i x}{F_i\left(\frac{L}{x}\right)} \right\}, \qquad (3.19)$$

where η_i represents the available offloading duration of MU i, and it can be recursively computed as follows:

$$\eta_i = \max \left\{ x - \sum_{i'=1}^{i-1} z_{i'}^{\text{bot}}(x), 0 \right\}, \qquad (3.20)$$

with the initial condition of $\eta_1 = x$. Besides, for each MU i with $N + 1 \leq i \leq I$, its unique optimal offloading duration is $z_i^{\text{bot}}(x) = 0$.

Proposition 3.1 means that we do not need to consider those MUs with $N + 1 \leq i \leq I$, since they are unhelpful for offloading the content. Thus, we define the following subset of the MUs, denoted by $\tilde{\mathcal{I}}$, as follows:

$$\tilde{\mathcal{I}} = \{i \,|\, i = 1, 2, ..., N\}. \tag{3.21}$$

The MUs in $\tilde{\mathcal{I}}$ are potentially helpful in terms of offloading the content. Notice that $\tilde{\mathcal{I}} = \emptyset$ if $N = 0$.

Using Proposition 3.1, we can express the optimal objective value of Problem (JTCOO-Bottom) as follows:

$$O_{\text{bot}}(x) = \sum_{i \in \tilde{\mathcal{I}}} M_i z_i^{\text{bot}}(x)(2^{\frac{L}{x}} - 1), \tag{3.22}$$

which will be used in the next section to solve Problem (JTCOO-Top).

3.4.2 Optimal Content Transmission: Solution of Top-Problem

Using $O_{\text{bot}}(x)$, we continue to solve Problem (JTCOO-Top) to determine the optimal transmission duration x^*.

$$\text{(JTCOO-Top):} \quad \max O_{\text{bot}}(x) + \alpha \left((2^{\frac{L}{x}} - 1) \frac{n}{\min_{i \in \mathcal{I}} \{g_{Bi}\}} + q_B \right) x + x \sum_{i \in \mathcal{I}} \beta_i h_i + \gamma x$$

$$\text{subject to:} \quad 0 \leq x \leq X^{\text{up}} = \left\{ \min_{i \in \mathcal{I}} \{ \frac{E_i^{\text{b}}}{h_i} \}, T^{\max} \right\},$$

$$\text{variables:} \quad x.$$

Directly solving (JTCOO-Top) is difficult, since x influences $\{z_i^{\text{bot}}(x)\}$ in (3.19), and consequently influences $O_{\text{bot}}(x)$ in (3.22) in a complicated fashion. Hence, we need to characterize different subregions for x, such that we can obtain the analytical form of $O_{\text{bot}}(x)$. The key idea of characterizing different subregions is that we will further identify those MUs (in set $\mathcal{I}$, i.e., the set of potential helpful MUs for performing offloading) that are *not eligible* to perform offloading under a given value of x, due to their limited transmit powers.

3.4.2.1 Characterizing Different Subregions for Variable x

Thresholds for excluding MUs not eligible for offloading data: We first consider the following threshold

$$\Gamma_i = \frac{L}{\log_2\left(1 + \frac{P_i^{\max}\min_{i' \neq i, i' \in \mathscr{I}}\{g_{ii'}\}}{n}\right)}, 1 \leq i \leq N, \qquad (3.23)$$

regarding whether MU i is an eligible candidate to perform offloading or not. Specifically, MU i is eligible for offloading, only if the transmission duration x satisfies $x \geq \Gamma_i$ (which leads to $F_i(L/x) \leq P_i^{\max}$). Otherwise, MU i is not eligible for offloading. Thus, starting from $x = \min\{\min_{i \in \mathscr{I}}\{E_i^b/h_i\}, T^{\max}\}$, a decrease in x means that the transmission rate of the content increases, and less MUs are eligible for performing offloading.

Based on the above consideration, we further reorder the thresholds defined in (3.23) in an ascending order as follows:

$$\Gamma_1 \leq \Gamma_2 \leq \dots \leq \Gamma_{l-1} \leq \Gamma_l \leq \Gamma_{l+1} \leq \dots \leq \Gamma_N. \qquad (3.24)$$

Different from (3.23), we now use subscript l as the index for thresholds $\{\Gamma_l\}_{1 \leq l \leq N}$ that follow the ordering in (3.24), i.e., $\Gamma_l \leq \Gamma_{l+1}$ always holds.

Moreover, given the index l of threshold Γ_l following (3.24), we define a mapping $T(l)$ to find the index of the MU that yields threshold Γ_l according to (3.23), i.e.,[1]

$$T(l) = \left\{s \in \tilde{\mathscr{I}} \mid \frac{\min_{i' \neq s, i' \in \mathscr{I}}\{g_{si'}\}P_s^{\max}}{n} = 2^{\frac{L}{\Gamma_l}} - 1\right\}. \qquad (3.25)$$

Recall that due to the reordering in (3.24), Γ_l does not correspond to MU l, and that is why we need to define the mapping $T(l)$ in (3.25). To clarify the ambiguity, we emphasize that in the rest of this chapter, subscript l is solely used as the index for thresholds $\{\Gamma_l\}_{1 \leq l \leq N}$.

Special MU on the "Boundary": Even if we focus on the subregion of $x \in [\Gamma_l, \Gamma_{l+1}]$, we still cannot analytically express the optimal solution of Problem (JTCOO-Bottom) $\{z_i^{bot}(x)\}_{i \in \mathscr{I}}$. We need to further consider the influence of x in constraint (3.15). The key step is to characterize a special MU (let us say MU v), such that the energy budget E_v^b of MU v is not used up, while the available offloading duration η_{v+1} given by (3.20) is zero. In particular, if such an MU v exists, then the BS does not need to perform broadcasting, since the MUs' offloading capabilities have not been fully utilized. Otherwise, the BS needs to perform broadcasting to finish delivering the whole content. Therefore, we consider the following two different types of cases:

- Type-I cases in which the BS does not need to perform broadcasting.
- Type-II cases in which the BS needs to perform broadcasting.

---

[1]Given index l, if there exist several different MUs in $\tilde{\mathscr{I}}$ such that condition in (3.25) is met, then $T(l)$ just represents such a set of MUs. This will not influence our following analytical results and the proposed algorithm. However, because of the randomness in different MUs' locations and the fading effect of the channel power gains, such case rarely happens.

Type-I cases that do not require the BS to perform broadcasting: The common property of the Type-I cases is as follows. Given x in the subregion $[\Gamma_l, \Gamma_{l+1}]$, there always exists a special MU $v \in \tilde{\mathscr{I}} \setminus \mathscr{J}_l$, such that the energy budget E_v^b of MU v is not used up, while the available offloading duration η_{v+1} given by (3.20) is zero. We denote this case by case (l, v), whose definition is as follows:

Definition 3.1 Case (l, v): Given that x in the subregion $[\Gamma_l, \Gamma_{l+1}]$, MU $v \in \tilde{\mathscr{I}} \setminus \mathscr{J}_l$ has its energy budget not used up, i.e., $E_v^b - E_v(x, z_v^{\text{bot}}(x)) > 0$ (where $E_v(x, z_v^{\text{bot}}(x))$ is given in (3.6)), while the available offloading duration η_{v+1} given by (3.20) is zero, i.e., $x = \sum_{i' \in \tilde{\mathscr{I}} \setminus \mathscr{J}_l}^{v} z_{i'}^{\text{bot}}(x)$.

There exist at most $N(N+1)/2$ such cases of Type-I. Given case (l, v), the optimal solution of Problem (JTCOO-Bottom), which is given in Proposition 3.1 before, can be further detailed as follows:

Proposition 3.2 *Given x and case (l, v), the optimal solution of Problem (JTCOO-Bottom) can be given by*

$$z_i^{\text{bot}}(x) = \frac{E_i^b - h_i x}{(2^{\frac{L}{x}} - 1)\frac{n}{\min_{i' \neq i, i' \in \mathscr{I}} \{g_{ii'}\}}}, \text{ when } 1 \leq i \leq v - 1, \text{ and } i \in \tilde{\mathscr{I}} \setminus \mathscr{J}_l, \quad (3.26)$$

$$z_v^{\text{bot}}(x) = x - \sum_{s=1, s \in \tilde{\mathscr{I}} \setminus \mathscr{J}_l}^{v-1} z_s^{\text{bot}}(x), \quad (3.27)$$

$$z_i^{\text{bot}}(x) = 0, \text{ when } v < i \leq I, \text{ or } i \in \mathscr{J}_l. \quad (3.28)$$

Type-II cases that require the BS to perform broadcasting: The common property of the Type-II cases is as follows. Given that x in the subregion $[\Gamma_l, \Gamma_{l+1}]$, each MU $i \in \tilde{\mathscr{I}} \setminus \mathscr{J}_l$ has used up its energy budget, while there still exists a nonzero available offloading duration, i.e., $\sum_{i \in \tilde{\mathscr{I}} \setminus \mathscr{J}_l} z_i^{\text{bot}}(x) < x$. This means that the BS needs to perform broadcasting to finish delivering the content. We denote this case by case (l, B) (where the capital letter "B" represents the BS), and its definition is as follows:

Definition 3.2 Case (l, B): Given x in subregion $[\Gamma_l, \Gamma_{l+1}]$, each MU $i \in \tilde{\mathscr{I}} \setminus \mathscr{J}_l$ has used up its energy budget, i.e., $E_i(x, z_i^{\text{bot}}(x)) = E_i^b$, and the BS still needs to perform broadcasting to finish delivering the content, i.e., $x - \sum_{i \in \tilde{\mathscr{I}} \setminus \mathscr{J}_l} z_i^{\text{bot}}(x) > 0$.

There exist $N + 1$ such cases of Type-II. Given case (l, B), the optimal solution of Problem (JTCOO-Bottom), which is given in Proposition 3.1 before, can be further detailed as follows:

Proposition 3.3 *Given x and case (l, B), the optimal solution of Problem (JTCOO-Bottom) can be given by*

$$z_i^{\text{bot}}(x) = \frac{E_i^b - h_i x}{(2^{\frac{L}{x}} - 1)\frac{n}{\min_{i' \neq i, i' \in \mathscr{I}} \{g_{ii'}\}}}, \text{ when } i \in \tilde{\mathscr{I}} \setminus \mathscr{J}_l, \quad (3.29)$$

$$z_i^{\text{bot}}(x) = 0, \text{ when } i \in \mathscr{I} \setminus \tilde{\mathscr{I}} \text{ or } i \in \mathscr{J}_l. \quad (3.30)$$

Correspondingly, in order to finish delivering the whole content, the BS broadcasts
for the duration which is equal to

$$x - \sum_{i \in \mathscr{F} \setminus \mathscr{I}_l} \frac{E_i^b - h_i x}{(2^{\frac{L}{x}} - 1) \frac{n}{\min_{i' \neq i, i' \in \mathscr{I}} \{g_{ii'}\}}}. \tag{3.31}$$

Until now, under a given x, we have analytically derived the optimal solution
in (3.26)–(3.28) of Problem (JTCOO-Bottom) in Proposition 3.2 by supposing that
case (l, v) (of Type-I) holds. Meanwhile, we also derive the corresponding optimal
solution in (3.29)–(3.31) of Problem (JTCOO-Bottom) in Proposition 3.3 by supposing that case (l, B) (of Type-II) holds. As a result, the optimal value of Problem
(JTCOO-Bottom), i.e., $O_{bot}(x)$ in (3.22), can be analytically detailed. We thus continue to solve Problem (JTCOO-Top) in the next two subsections, in which we will
also provide the conditions to verify whether case (l, v) (or case (l, B)) holds or not.

3.4.2.2 Analytical Optimal Content Transmission Duration for Each Case (l, v)

Given x and case (l, v), we introduce function $W_{l,v}(x)$ to denote the objective function
of Problem (JTCOO-Top) under case (l, v). By substituting (3.26) and (3.27) into
(3.22), we can compactly express function $W_{l,v}(x)$ as follows:

$$W_{l,v}(x) = (2^{\frac{L}{x}} - 1)x \left(\alpha \frac{n}{g_{Bv}} + \beta_v \frac{n}{\min_{i \neq v, i \in \mathscr{I}} \{g_{vi}\}} \right)$$

$$+ x \left(\alpha q_B + \sum_{i \in \mathscr{I}} \beta_i h_i + \gamma + S_{l,v} \right) + Q_{l,v}, \tag{3.32}$$

where both $S_{l,v}$ and $Q_{l,v}$ are constant and depend on case (l, v):

$$S_{l,v} = \sum_{i=1, i \in \mathscr{F} \setminus \mathscr{I}_l}^{v-1} \left(\alpha \left(\frac{1}{g_{Bv}} - \frac{1}{g_{Bi}} \right) + \beta_v \frac{1}{\min_{i \neq v, i' \in \mathscr{I}} \{g_{vi'}\}} \right.$$

$$\left. - \beta_i \frac{1}{\min_{i' \neq i, i' \in \mathscr{I}} \{g_{ii'}\}} \right) h_i \min_{i' \neq i, i' \in \mathscr{I}} \{g_{ii'}\}, \tag{3.33}$$

$$Q_{l,v} = \sum_{i=1, i \in \mathscr{F} \setminus \mathscr{I}_l}^{v-1} \left(\alpha \left(\frac{1}{g_{Bi}} - \frac{1}{g_{Bv}} \right) + \beta_i \frac{1}{\min_{i' \neq i, i' \in \mathscr{I}} \{g_{ii'}\}} \right.$$

$$\left. - \beta_v \frac{1}{\min_{i' \neq v, i' \in \mathscr{I}} \{g_{vi'}\}} \right) E_i^b \min_{i' \neq i, i' \in \mathscr{I}} \{g_{ii'}\}. \tag{3.34}$$

Therefore, given case (l, v), solving Problem (JTCOO-Top) becomes equivalent to solving

$$\text{(JTCOO-Top-(l,v)):} \quad \min_{x} W_{l,v}(x),$$

$$\text{subject to:} \quad \Gamma_l \leq x \leq \min\left\{X^{\text{up}}, \Gamma_{l+1}\right\},$$

$$\text{variables:} \quad x,$$

where X^{up} has been defined in (3.16).

Let $x_{l,v}^*$ denote the optimal solution of Problem (JTCOO-Top-(l,v)). Although $W_{l,v}(x)$ is complicated, we can analytically derive $x_{l,v}^*$ in the following proposition:

Proposition 3.4 *Given case (l, v), the optimal solution for Problem (JTCOO-Top-(l,v)) can be given by*

$$x_{l,v}^* = \left[(\ln 2)\frac{L}{1 + \mathcal{W}\left(e^{-1}(\frac{B_{l,v}}{A_{l,v}} - 1)\right)}\right]_{\Gamma_l}^{\min\{X^{\text{up}}, \Gamma_{l+1}\}}, \tag{3.35}$$

where expression $[x]_a^b = \min\{\max\{a, x\}, b\}$, and $\mathcal{W}(.)$ represents the Lambert-W function [29], i.e., the inverse function of $f(w) = w\exp(w)$. Meanwhile, parameters $A_{l,v}$ and $B_{l,v}$ are given by

$$A_{l,v} = \alpha\frac{n}{g_{Bv}} + \beta_v\frac{n}{\min_{i \neq v, i \in \mathscr{I}}\{g_{vi}\}}, \tag{3.36}$$

$$B_{l,v} = \alpha q_B + \sum_{i \in \mathscr{I}} \beta_i h_i + \gamma + S_{l,v}. \tag{3.37}$$

Accordingly, $W_{l,v}^*(x_{l,v}^*) = A_{l,v}x_{l,v}^*(2^{\frac{L}{x_{l,v}^*}} - 1) + B_{l,v}x_{l,v}^* + Q_{l,v}$.

The result in Proposition 3.4 is based on the assumption that case (l, v) holds. We thus need to use the derived $x_{l,v}^*$ in Proposition 3.4 to verify whether case (l, v) holds or not. This leads to the following proposition:

Proposition 3.5 (Validation of case (l, v)) *Case (l, v) holds, if the derived $x_{l,v}^*$ in (3.35) meets the following two conditions:*

$$x_{l,v}^*(2^{\frac{L}{x_{l,v}^*}} - 1) + x_{l,v}^* \sum_{i=1, i \in \mathscr{\bar{I}} \backslash \mathscr{I}_l}^{v-1} \frac{h_i}{n} \min_{i' \neq i, i' \in \mathscr{I}}\{g_{ii'}\}$$

$$\geq \sum_{i=1, i \in \mathscr{\bar{I}} \backslash \mathscr{I}_l}^{v-1} \frac{E_i^b}{n} \min_{i' \neq i, i' \in \mathscr{I}}\{g_{ii'}\}, \tag{3.38}$$

$$x_{l,v}^*(2^{\frac{L}{x_{l,v}^*}} - 1) + x_{l,v}^* \sum_{i=1, i \in \widetilde{\mathcal{F}} \backslash \mathcal{J}_l}^{v} \frac{h_i}{n} \min_{i' \neq i, i' \in \mathcal{J}} \{g_{ii'}\}$$

$$\leq \sum_{i=1, i \in \widetilde{\mathcal{F}} \backslash \mathcal{J}_l}^{v} \frac{E_i^b}{n} \min_{i' \neq i, i' \in \mathcal{J}} \{g_{ii'}\}. \qquad (3.39)$$

3.4.2.3 Analytical Optimal Content Transmission Duration for Each Case (l, B)

Given x and case (l, B), we introduce $W_{l,B}(x)$ to denote the objective function of Problem (JTCOO-Top). By substituting (3.29) and (3.31) into (3.22), we can compactly express function $W_{l,B}(x)$ as follows:

$$W_{l,B}(x) = (2^{\frac{L}{x}} - 1)x\alpha \frac{n}{\min_{i \in \mathcal{J}} \{g_{Bi}\}} + x\left(\alpha q_B + \sum_{i \in \mathcal{J}} \beta_i h_i + \gamma + S_{l,B}\right)$$

$$+ Q_{l,B}, \qquad (3.40)$$

where both $S_{l,B}$ and $Q_{l,B}$ are constant and depend on index l:

$$S_{l,B} = -\sum_{i \in \widetilde{\mathcal{F}} \backslash \mathcal{J}_l} \left(\alpha \frac{1}{g_{Bi}} + \beta_i \frac{1}{\min_{i' \neq i, i' \in \mathcal{J}} \{g_{ii'}\}}\right.$$

$$\left. -\alpha \frac{1}{\min_{i' \in \mathcal{J}} \{g_{Bi'}\}}\right) h_i \min_{i' \neq i, i' \in \mathcal{J}} \{g_{ii'}\}, \qquad (3.41)$$

$$Q_{l,B} = \sum_{i \in \widetilde{\mathcal{F}} \backslash \mathcal{J}_l} \left(\alpha \frac{1}{g_{Bi}} + \beta_i \frac{1}{\min_{i' \neq i, i' \in \mathcal{J}} \{g_{ii'}\}}\right.$$

$$\left. -\alpha \frac{1}{\min_{i' \in \mathcal{J}} \{g_{Bi'}\}}\right) E_i^b \min_{i' \neq i, i' \in \mathcal{J}} \{g_{ii'}\}. \qquad (3.42)$$

Given case (l, B), solving Problem (JTCOO-Top) becomes equivalent to solving

(JTCOO-Top-(l,B)): $\min_{x} W_{l,B}(x),$

subject to: $\Gamma_l \leq x \leq \min\{X^{\mathrm{up}}, \Gamma_{l+1}\},$

variables: $x.$

Let $x_{l,B}^*$ denote the optimal solution of Problem (JTCOO-Top-(l,B)). We can analytically derive $x_{l,B}^*$ in the following proposition.

Proposition 3.6 *Given case (l, B), the optimal solution of Problem (JTCOO-Top-(l,B)) can be given by*

$$x_{l,B}^* = \left[(\ln 2) \frac{L}{1 + \mathcal{W} \left(e^{-1} \left(\frac{B_{l,B}}{A_{l,B}} - 1 \right) \right)} \right]_{\Gamma_l}^{\min\{X^{up}, \Gamma_{l+1}\}}, \tag{3.43}$$

where parameters $A_{l,B}$ and $B_{l,B}$ are, respectively, given by

$$A_{l,B} = \alpha \frac{n}{\min_{i \in \mathscr{I}} \{g_{Bi}\}}, \tag{3.44}$$

$$B_{l,B} = \alpha q_B + \sum_{i \in \mathscr{I}} \beta_i h_i + \gamma + S_{l,B}. \tag{3.45}$$

Accordingly, $W_{l,B}^*(x_{l,B}^*) = A_{l,B} x_{l,B}^* (2^{\frac{L}{x_{l,B}^*}} - 1) + B_{l,B} x_{l,B}^* + Q_{l,B}.$

The result in Proposition 3.6 is based on the assumption that case (l, B) holds. We thus need to use the derived $x_{l,B}^*$ in Proposition 3.6 to verify whether case (l, B) holds or not. This leads to the following Proposition.

Proposition 3.7 (Validation of case (l,B)) *Case (l,B) holds, if the derived $x_{l,B}^*$ in (3.43) meets the following conditions:*

$$x_{l,B}^* (2^{\frac{L}{x_{l,B}^*}} - 1) + x_{l,B}^* \sum_{i \in \mathscr{I} \setminus \mathscr{I}_l} \frac{h_i}{n} \min_{i' \neq i, i' \in \mathscr{I}} \{g_{ii'}\} \geq \sum_{i \in \mathscr{I} \setminus \mathscr{I}_l} \frac{E_i^b}{n} \min_{i' \neq i, i' \in \mathscr{I}} \{g_{ii'}\}. \tag{3.46}$$

3.4.3 Algorithm for Optimal Offloading Solution

Using the above analytical results, we propose a Joint Optimization of Transmission and Offloading Durations (JOTOD) algorithm to compute the optimal transmission duration for the whole content and each MU's offloading duration for Problem (JTCOO-Top).

Algorithm JOTOD enumerates all Type-I cases and Type-II cases. To this end, it consists of a two-layered loop, i.e., (i) an *outer While loop* from Step 3 to Step 25 for enumerating all possible index l, and (ii) given index l, an *inner While loop* from Step 5 to Step 15 for enumerating index v such that each possible case (l, v) is evaluated, and moreover, the additional steps from Step 16 to Step 23 for evaluating case (l, B). Specifically, for each enumerated case (l, v), we derive $x_{l,v}^*$ based on Proposition 3.4 in Step 6, and further verify whether case (l, v) holds in Step 7. If case (l, v) is valid and the obtained $W_{l,v}(x_{l,v}^*)$ can improve the currently best value ϕ, then we update ϕ and record the currently best solution of Problem (JTCOO) in Step 11. Similarly, for each enumerated case (l, B), we derive $x_{l,B}^*$ based on Proposition 3.6 in Step 16, and further verify whether case (l, B) holds in Step 17. If case (l, B) is valid and the obtained $W_{l,B}(x_{l,B}^*)$ can improve the currently best value ϕ, then we update ϕ and record the currently best solution of Problem (JTCOO) in Step 21. Finally, Algorithm JOTOD outputs the optimal solution of Problem (JTCOO) in Step 26 based on the currently best solution. We thus complete solving Problem (JTCOO).

Algorithm JOTOD: to compute the optimal solution for Problem (JTCOO)

1: Initialize ϕ as a very large positive number, e.g., $\phi = 10^8$.
2: Set $l = N$, where N is obtained from the ordering (3.18).
3: **while** $l \geq 0$ **do**
4:     Set $v = 1$.
5:     **while** $v \leq N$ and $v \notin \mathcal{J}_l$ **do**
6:       Derive $x_{l,v}^*$ according to (3.35).
7:       **if** $x_{l,v}^*$ meets (3.38) and (3.39) simultaneously **then**
8:         Evaluate $W_{l,v}(x_{l,v}^*)$ according to (3.32).
9:         **if** $W_{l,v}(x_{l,v}^*) < \phi$ **then**
10:           Derive $\{z_j^{\mathrm{bot}}(x_{l,v}^*)\}_{j \in \mathcal{J}}$ according to (3.26)–(3.28).
11:           Update $\phi = W_{l,v}(x_{l,v}^*)$, and record the currently best solution of Problem (JTCOO) as:

$$x^{*,c} = x_{l,v}^*, r^{*,c} = \frac{L}{x_{l,v}^*}, \text{ and } z_j^{*,c} = z_j^{\mathrm{bot}}(x_{l,v}^*), \forall j \in \mathcal{J}$$

12:         **end if**
13:       **end if**
14:       Set $v = v + 1$.
15:     **end while**
16:     Derive $x_{l,\mathrm{B}}^*$ according to (3.43).
17:     **if** $x_{l,\mathrm{B}}^*$ meets (3.46) **then**
18:       Evaluate $W_{l,\mathrm{B}}(x_{l,\mathrm{B}}^*)$ according to (3.40).
19:       **if** $W_{l,\mathrm{B}}(x_{l,\mathrm{B}}^*) < \phi$ **then**
20:         Derive $\{z_j^{\mathrm{bot}}(x_{l,\mathrm{B}}^*)\}_{j \in \mathcal{J}}$ according to (3.29) and (3.30).
21:         Update $\phi = W_{l,\mathrm{B}}(x_{l,\mathrm{B}}^*)$, and record the currently best solution of Problem (JTCOO) as:

$$x^{*,c} = x_{l,\mathrm{B}}^*, r^{*,c} = \frac{L}{x_{l,\mathrm{B}}^*}, \text{ and } z_j^{*,c} = z_j^{\mathrm{bot}}(x_{l,\mathrm{B}}^*), \forall j \in \mathcal{J}.$$

22:       **end if**
23:     **end if**
24:     Set $l = l - 1$.
25: **end while**
26: Set the optimal solution of Problem (JTCOO) as: $x^* = x^{*,c}, r^* = r^{*,c}$, and $z_j^* = z_j^{*,c}, \forall j \in \mathcal{J}$.

3.5 Numerical Results

In this section, we perform numerical simulations to validate Algorithm JOTOD and the performance achieved by the MUs' optimal cooperations. We setup a scenario as shown in Fig. 3.2, in which the BS is located at the origin $(0, 0)$. The group of MUs are randomly and independently located (according to a uniform distribution) within a circle. The central of the circle is $(D, 0)$, and its radius is R. We set $D = 50\,\mathrm{m}$ and $R = 5\,\mathrm{m}$ at the beginning (but will vary D and R later on). In particular, we assume that the MUs do not move during the period of interest, e.g., one period of $T^{\max}$ (i.e., the delay bound for finishing delivery of the content). Thus, the channel power gain from the BS to each MU and that between the MUs remain unchanged (e.g., within one period of $T^{\max}$). In particular, we model the channel power gain from the BS to each MU i as $g_{\mathrm{B}i} = \frac{\xi_{\mathrm{B}i}}{l_{\mathrm{B}i}^\kappa}$, where parameter $l_{\mathrm{B}i}$ denotes the distance between the BS and MU i, parameter κ denotes the power scaling factor for the path loss, and parameter $\xi_{\mathrm{B}i}$ follows an exponential distribution with unit mean for capturing the fading.

Fig. 3.2 Network scenario used for numerical experiments. We use $\mathscr{I} = \{1, 2, 3, 4, 5\}$ as an example. The MUs are randomly located within a *circle*. The central of the *circle* is $(D, 0)$. The radius of the circle is R

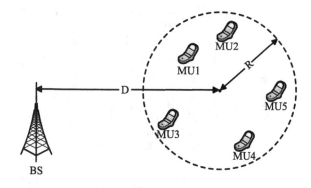

Fig. 3.3 Accuracy of Algorithm JOTOD in solving Problem (JTCOO). We vary the number of the MUs $I = 10, 20, ..., 50$ and the distance $D = 20, 40, 60$. For each tested case, the result (i.e., the total system cost) is averaged over 200 randomly generated network scenarios

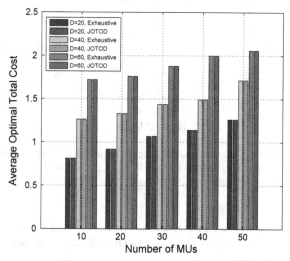

Similar to [22], we set the static circuit power consumption of the BS during data transmission as $q_B = 1\,\text{W}$, the circuit power consumption of MU i during data reception as $h_i = 0.01\,\text{W}$ (i.e., 1% of the static circuit power consumption of the BS). Besides, the maximum transmit power of each MU is $P_i^{\max} = 0.1\,\text{W}$, and the energy budget of each MU is $E_i^b = 0.1\,\text{J}$. The channel bandwidth is $1\,\text{MHz}$ for the cellular link and the link between different MUs. The size of the content is $L = 1\,\text{Mbits}$. Besides, we set $\alpha = 1$, $\beta_i = 2$, $\forall i \in \mathscr{I}$, and $\gamma = 1$.

Accuracy of Proposed Algorithm: Figure 3.3 validates the accuracy of Algorithm JOTOD in solving Problem (JTCOO) optimally. We vary the number of the MUs $I = 10, 20, ..., 50$ and the distance $D = 20, 40, 60$. For each tested case, we plot the average result (i.e., the total system cost) over 200 network scenarios which are randomly generated as described earlier. Figure 3.3 shows that Algorithm JOTOD achieves the optimal total system cost which is exactly same as the global optimum

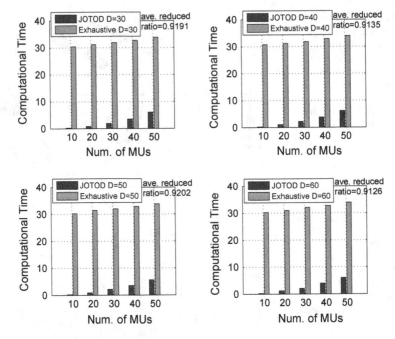

Fig. 3.4 Computational Efficiency of Algorithm JOTOD. We vary the distance $D = 30, 40, 50, 60$ and the number of the MUs $I = 10, 20, ..., 50$. For each tested case, the result (i.e., the computational time) is averaged over 200 randomly generated network scenarios

found by the exhaustive search method,[2] thus validating the accuracy of Algorithm JOTOD. Besides, it is observed that the total system cost increases in the number of the MUs.

Computational Efficiency of Proposed Algorithm: Figure 3.4 validates the computational efficiency of Algorithm JOTOD. Specifically, we vary the distance $D = 30, 40, 50, 60$ and the number of the MUs $I = 10, 20, ..., 50$. For each tested case, we plot the average result (i.e., the computational time) over 200 randomly generated network scenarios. Figure 3.4 shows that Algorithm JOTOD consumes a significantly less computational time than the exhaustive search method. Specifically, for each tested distance D, Algorithm JOTOD reduces the computational time by more than 90% on average. Furthermore, by comparing different subplots in Fig. 3.4, we can observe that the computational time of Algorithm JOTOD increases mainly as the number of the MUs increases, but varies slightly as the distance D changes. This

---

[2]The exhaustive search method enumerates the transmission duration x using a very small step size. For each enumerated x, we again use (3.19) to determine $\{z_j^{bot}(x)\}$ and thus evaluate $O_{bot}(x)$. Therefore, the exhaustive search method is guaranteed to achieve the global optimum for Problem (JTCOO) with a negligible loss, as long as the chosen step size is small enough. However, the downside of the exhaustive search method is that it consumes a significant computational time.

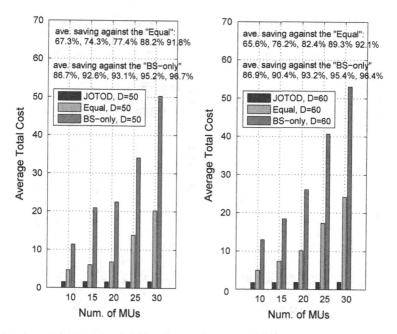

Fig. 3.5 Advantage of saving the total system cost using the optimal D2D-assisted traffic offloading. We plot the average total system versus different number of the MUs $I = 10, 15, ..., 30$. *Left subplot* the distance between the BS and the central of the circle $D = 50$. *Right subplot* $D = 60$. For each tested case, the result (i.e., the total system cost) is averaged over 200 randomly generated network scenarios. The numbers illustrated on the top of each subplot denote the average saving ratios of the total system cost using the optimal offloading solution against the Equal-distribution approach and the BS-only approach, respectively

result is consistent with our previous discussions about the computational complexity of Algorithm JOTOD.

Advantage of the Optimal D2D-assisted Offloading Solution: We present the advantage of reducing the total system cost using the optimal D2D-assisted offloading solution in Figs. 3.5 and 3.6. To show this advantage, we compare the result of the optimal offloading solution (i.e., the result of Algorithm JOTOD) with two other heuristic approaches, namely, the BS-only approach and the Equal-division approach. In the BS-only approach, the BS directly broadcasts the whole content to all MUs and only optimizes its transmission duration to minimize the total system cost. In the Equal-division approach, all the helpful MUs in $\tilde{\mathscr{I}}$ equally share the transmission duration for offloading the content, i.e., $z_i = x/N$, for all MUs in $\tilde{\mathscr{I}}$, and the BS optimizes its transmission duration accordingly. Notice that if some MU, e.g., MU i, cannot afford the required transmit power or the required offloading duration, then the BS will take over the job of MU i to deliver the content by broadcasting.

In Fig. 3.5, we consider the distance between the BS and the central of circle $D = 50$ (in the left subplot) and $D = 60$ (in the right subplot), and vary the number of the MUs $I = 10, 15, ..., 30$. Figure 3.5 shows that the optimal

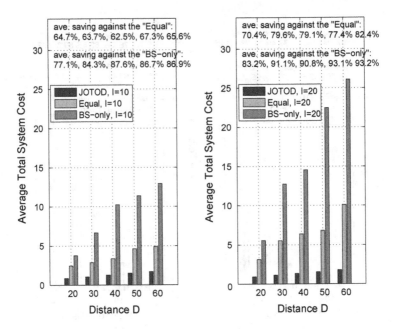

Fig. 3.6 Advantage of optimal D2D-assisted offloading in saving the total system cost. We plot the average total system versus different distance $D = 20, 30, ..., 60$. *Left subplot* We fix the number of MUs $I = 10$. *Right subplot* $I = 20$. For each tested case, the result (i.e., the total system cost) is averaged over 200 randomly generated network scenarios. Besides, the numbers illustrated on the top of each subplot denote the average saving ratios of the total system cost using the optimal offloading solution against the Equal-distribution approach and the BS-only approach, respectively

D2D-assisted offloading solution significantly outperforms the BS-only approach and the Equal-distribution approach in terms of lowering the total system cost. For each tested case, we mark out the average saving ratios of the total system cost (i.e., the numbers listed on the top of subplot) using optimal offloading solution against the Equal-distribution approach and the BS-only approach, respectively. Remarkably, the optimal D2D-assisted offloading can save more than 60% of the system cost compared to the Equal-distribution approach, and saving more than 70% of the system cost compared to the BS-only approach. Moreover, the results show that the average saving ratio increases in the number of the MUs, i.e., a larger portion of the system cost is reduced. This is because a larger number of the MUs provides a larger freedom in performing offloading, which consequently yields a larger gain in terms of lowering the total system cost.

In Fig. 3.6, we consider the number of MUs $I = 10$ (in the left subplot) and $I = 20$ (in the right subplot), and vary the distance $D = 20, 30, ..., 60$. For each tested case, we mark out the average saving ratios of the total system cost (i.e., the numbers listed on the top of each subplot) using the optimal offloading solution. Figure 3.6 again shows that the optimal offloading solution can significantly reduce the total system cost compared with the Equal-distribution approach and the BS-only

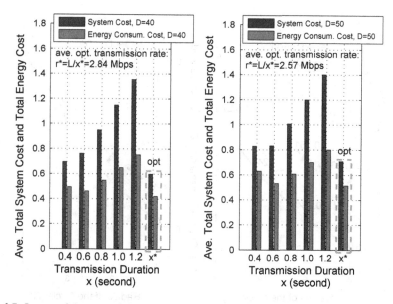

Fig. 3.7 Impact of the transmission duration x on the total system cost and total cost for energy consumption. Each result represents the average result over 200 randomly generated network scenarios. We fix the number of MUs $I = 10$ and $\alpha = \beta_i = \gamma = 0.5$, $\forall i \in \mathscr{I}$. *Left subplot* $D = 40$ m. *Right subplot* $D = 50$ m

approach. Meanwhile, the comparisons between the two subplots also verify that a larger saving ratio can be achieved when more MUs coexist for offloading.

To further evaluate the advantage of the optimal offloading solution that jointly considers the content transmission rate and the MUs' offloading durations, we compare the optimal cooperative scheme with another heuristic scheme with the fixed content transmission rate in Fig. 3.7. Specifically, in the heuristic scheme, the content transmission duration x is heuristically fixed (which corresponds to a heuristically chosen transmission rate r), but the MUs' offloading durations are optimally calculated according to Proposition 3.1. Each result in Fig. 3.7 represents the average result for 200 randomly generated network scenarios.

In the left subplot of Fig. 3.7 (with $D = 40$ m), the rightmost result labeled with x^* denotes the output of the optimal offloading solution. For each randomly generated scenario, the corresponding x^* is different. That is why we label the result with x^*, instead of a particular numerical value. Meanwhile, the other five results labeled with $x = 0.4, 0.6, 0.8, 1.0$, and 1.2 denote the output of the heuristic scheme with the content transmission duration fixed at $x = 0.4, 0.6, 0.8, 1.0$, and 1.2 s (which correspond to that the transmission rate $r = 2.5, 1.67, 1.25, 1.0, 0.83$ Mbps), respectively. As shown in the left subplot of Fig. 3.7, the optimal offloading solution can effectively reduce the total system cost as well as total cost for energy consumption, compared with the heuristic scheme with the fixed transmission durations. In particular, as we have marked out in the left subplot, the average optimal transmission

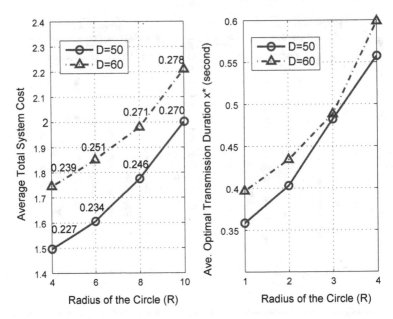

Fig. 3.8 Impact of the MUs' distribution on the optimal total system cost and the optimal transmission duration x^* (yielded by the optimal offloading solution). *Left subplot* the total system cost versus different (D, R). For each tested case, the result (i.e., the total system cost) is averaged over 200 randomly generated network scenarios. The number illustrated above each tested case represents the ratio between the cellular-link usage cost and the total system cost. *Right subplot* the optimal transmission duration x^* versus different (D, R)

rate is 2.84 Mbps. The right subplot of Fig. 3.7 (with $D = 50$ m) shows the similar advantage of the optimal offloading solution, with the average optimal transmission rate equal to 2.57 Mbps. The results in Fig. 3.7 again verify the importance of jointly optimizing the content transmission rate and the MUs' offloading durations.

Impact of MUs' Geographic Distribution: We evaluate the impact of the MUs' geographical distribution on the system performance by varying the distance D and the radius R. Recall that the tuple of (D, R) locates the circle in which the MUs are randomly distributed. The left subplot of Fig. 3.8 plots the total system cost (yielded by the optimal offloading solution) under different D and R, with the total number of MUs $I = 20$. The left subplot of Fig. 3.8 shows that the total system cost increases in R, since a larger geographical distribution of the MUs necessitates a larger transmit powers of the MUs for performing offloading and thus yields a greater system cost. For the similar reason, the total system cost also increases in D, since the BS needs a larger transmit power to transmit to some selected MUs (this point also has been reflected in Figs. 3.3, 3.5, and 3.6). Moreover, in the left subplot, we use the optimal offloading solution to compute the ratio between the cellular-link usage cost and the total system cost, and mark out this ratio for each tested case. Interestingly, the results show that the ratio increases in R, which means that the cellular-link usage cost tends to be more significant in the total system cost. The trend is reflected in the right subplot of Fig. 3.8 which shows the optimal transmission duration x^* versus different values of radius R, with the parameter settings corresponding to the left subplot. The results show the optimal transmission duration x^* also increases in R. This is because a larger R means a larger transmit power required by each MU to perform offloading, while prolonging transmission duration can reduce the content transmission rate and thus reduce the required transmit powers.

3.6 Extension for Distributing Multiple Pieces of Contents

3.6.1 System Model and Problem Formulation

We further extend the proposed MUs' D2D-assisted offloading for distributing multiple different contents, and different MUs are interested in downloading different subsets of contents. In this extension, we consider a group of contents, denoted by $\mathcal{K} = \{1, 2, ..., K\}$, available at the BS. We use L^k to denote the size of content k. Each MU i is interested to get a subset of $\mathcal{K}$. We use a U-by-K matrix $\mathbf{A}$ to denote the relationship between the MUs and the contents, i.e., $A_{i,k} = 1$ if MU i is interested in getting content k, and $A_{i,k} = 0$ otherwise. We use $\mathcal{C}_i$ to denote the set of the interested contents which MU i wants to receive, i.e., $\mathcal{C}_i = \{k \in \mathcal{K} | A_{i,k} = 1, i \in \mathcal{I}\}$, and we use Ω^k to denote the set of MUs which are interested in obtaining content k, i.e., $\Omega^k = \{i \in \mathcal{I} | A_{i,k} = 1, k \in \mathcal{K}\}$. In the multiple-content scenario, the BS determines the transmission rate r^k of content k, which is equivalent to its transmission duration $x^k = \frac{L^k}{r^k}$. The BS first sends the data of content k to some selected MU i, and MU i offloads its received data via broadcasting to the other MUs in Ω^k. Let z_i^k denote the offloading duration of MU $i \in \Omega^k$ for content k. There exists $\sum_{i \in \Omega^k} z_i^k \leq x^k$, i.e., the total offloading duration of all MUs in Ω^k cannot exceed x^k. Figure 3.9a shows the network scenario in which MU 1, MU 4, MU 6 are in Ω^1, and MU 2, MU 3, MU 4, MU 5 are in Ω^2, and the interest matrix $\mathbf{A}$ in this scenario is given by the table. Figures 3.9b–d illustrate the offloading process for content 1 step by step. Figures 3.9b and c show the cases of offloading through MU 1 and MU 4, respectively. Each of them sends their received data to other MUs in Ω^1. Finally, taking into account the MUs' limited transmit powers and energy capacities, the BS is allowed to directly broadcast content 1 to all MUs in Ω^1 for finishing the delivery of content 1. The corresponding duration for the BS to perform broadcasting is $x^1 - \sum_{i \in \Omega^1} z_i^1$. Figure 3.9d shows the case of BS's broadcasting to all MUs. However, such a broadcasting consumes the BS a significant transmit power, since the BS has to take into account the MUs with the worst channel condition from it. In

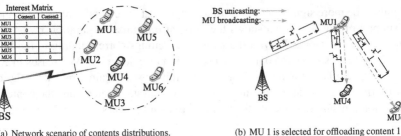

(a) Network scenario of contents distributions.

(b) MU 1 is selected for offloading content 1 for duration z_1^1.

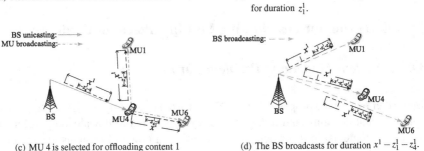

(c) MU 4 is selected for offloading content 1 for duration z_4^1.

(d) The BS broadcasts for duration $x^1 - z_1^1 - z_4^1$.

Fig. 3.9 Illustration of system model. Subplot **a** network scenario of content distributions. Subplot **b** MU 1 is selected for offloading content 1 for duration z_1^1. Subplot **c** MU 4 is selected for offloading content 1 for duration z_4^1. Subplot **d** The BS broadcasts for duration $x^1 - z_1^1 - z_4^1$

this case, the optimal design of MUs' D2D-assisted offloading turns to how to properly control different contents' transmissions and the MUs' offloading duration for different contents of interests. Such a design is much more challenging than that in Sect. 3.3, since we need to further take into account how each MU properly allocates its limited energy capacity for different contents of its interest.

With the similar modelings discussed in Sect. 3.2.2, we can further make the following modelings. First, the total energy consumption of the BS for delivering content k is:

$$E_{BS}^k = E_{B,\text{ui}}^k + E_{B,\text{bc}}^k = \sum_{i \in \Omega^k} \left(F_{Bi}^k(r^k) + q_B \right) z_i^k + \left(F_{B0}^k(r^k) + q_B \right) \left(x^k - \sum_{i \in \Omega^k} z_i^k \right),$$

$$(3.47)$$

where the total energy consumption (E_{BS}^k) is composed by the energy consumption for unicasting and that for broadcasting. Meanwhile, MU i's energy consumption for content k is

Table 3.2 Important Parameters and Variables

$\mathcal{K}$	The set of contents $\mathcal{K} = \{1, 2, ..., K\}$	T	The upper bound of time for transmitting all contents
W_i	The bandwidth of MU i	$\mathbf{A}$	The relationship between the MUs and the contents
W_B	The bandwidth of the BS	$E^k_{B,ui}$	The energy consumption of the BS for unicasting content k to selected MUs
Ω^k	The set of MUs which are interested in obtaining content k	$E^k_{B,bc}$	The energy consumption of the BS for broadcasting content k to all MUs who are interested in
$\mathcal{C}_i$	The set of the interested contents which MU i wants to receive	E^k_{BS}	The total energy consumption of the BS for content k
$E^k_{i,tx}$	The energy consumption of each MU i for offloading its received parts of content k	$E^k_{i,tot}$	The total energy consumption of each MU i for content k
$E^k_{i,rx}$	The energy consumption of each MU i for receiving content k		

$$E^k_{i,\text{tot}} = E^k_{i,\text{tx}} + E^k_{i,\text{rx}} = \left(F^k_i(r^k) + q_i \right) z^k_i + h_i x^k, \qquad (3.48)$$

where MU i's energy consumption is composed by the energy consumption of each MU i for offloading and that for receiving content k.

We formulate a Joint Transmission Control and Offloading Optimization problem for Multiple contents (i.e., Problem (JTCOO-M)). Problem (JTCOO-M) jointly controls each content k's transmission rate r^k (which is equivalent to its transmission duration x^k as reflected in (3.50)) and each MU i's offloading duration $\{z^k_i\}_{k \in \mathcal{C}_i}$ for its interested contents. Table 3.2 lists all important symbols used in the formulation. The objective is to minimize the total energy consumption of the BS and all MUs in $\mathcal{I}$. Constraint (3.51) guarantees that the total offloading duration of all MUs in Ω^k cannot exceed content k's transmission duration x^k. Constraint (3.52) ensures that the aggregate transmission duration of all contents in $\mathcal{K}$ cannot exceed an upper bound T, which can be considered as the length of a time slot. Constraint (3.53) ensures that the total energy consumption of MU i, including its energy consumption for offloading contents and its energy consumption for content reception, cannot exceed its energy budget E^b_i. Constraint (3.54) means that MU i can be selected for offloading content k only if the required transmit power plus the fixed circuit power is below its power capacity $P^{\max}_i$.

(JTCOO-M) : $\min V(\{r^k\}_{\forall k \in \mathcal{K}}, \{x^k\}_{\forall k \in \mathcal{K}}, \{z_i^k\}_{\forall i \in \mathcal{I}, k \in \mathcal{K}}) = \sum_{k \in \mathcal{K}} E_{BS}^k$

$$+ \sum_{i \in \mathcal{I}} \sum_{k \in \mathcal{C}_i} E_{i,\text{tot}}^k \quad (3.49)$$

subject to: $x^k r^k = L^k, \forall k \in \mathcal{K},$ \hfill (3.50)

$$\sum_{i \in \Omega^k} z_i^k \leq x^k \forall k \in \mathcal{K},$$ \hfill (3.51)

$$\sum_{k \in \mathcal{K}} x^k \leq T,$$ \hfill (3.52)

$$\sum_{k \in \mathcal{C}_i} \left(F_i^k(r^k) + q_i \right) z_i^k + \sum_{k \in \mathcal{C}_i} h_i x^k \leq E_i^{\text{b}}, \forall i \in \mathcal{I},$$ \hfill (3.53)

$$z_i^k \leq x^k \cdot \mathbb{I}\left(F_i^k(r^k) + q_i \leq P_i^{\text{max}} \right), \forall i \in \mathcal{I}, k \in \mathcal{K},$$ \hfill (3.54)

$$z_i^k \geq 0, \forall i \in \mathcal{I}, k \in \mathcal{K},$$ \hfill (3.55)

variables: $\{r^k\}_{\forall k \in \mathcal{K}}, \{x^k\}_{\forall k \in \mathcal{K}}, \{z_i^k\}_{\forall i \in \mathcal{I}, k \in \mathcal{K}}.$

3.6.2 Illustrative Approach for Optimal Solution

There exist two types of coupling effects in Problem (JTCOO-M). First, subject to each MU i's limited transmit power and energy budget, the optimization of the transmission rate and that of each MU i's offloading duration are strongly coupled. According to (3.54), a shorter transmission duration x^k of content k requires a greater transmit power. As a result, MU i with a low transmit power capacity P_i^{max} might become ineligible to perform offloading if (3.54) fails to hold. Moreover, according to (3.53), a greater x^k yields that MU $i \in \Omega^k$ can only perform offload for content k for a shorter duration. Second, since different MUs in Ω^k are cooperative in offloading content k, their respective offloading durations are subject to (3.51).

Due to the above two coupling effects and the discrete set of transmission rate of contents, Problem (JTCOO-M) is a complicated nonconvex optimization problem, which is difficult to solve. Despite this difficulty, we solve Problem (JTCOO-M) by exploiting the layered structure below. Suppose that the transmission rates $\{r^k\}_{k \in \mathcal{K}}$ of different contents are given in advance, which is equivalent to that the corresponding transmission duration $\{x^k\}_{k \in \mathcal{K}}$ of different contents are given. Thus, a close look at Problem (JTCOO-M) shows that it can be decomposed into two subproblems, namely, a top optimization problem that optimizes the transmission rate of each content, and a corresponding bottom optimization problem that optimizes the offloading durations of each MU under the given transmission rates of different contents. Figure 3.10 shows relationship between the top and bottom problems. A key advantage of the above decomposition is that the bottom problem is a linear programming problem regarding to $\{z_i^k\}_{\forall i \in \mathcal{I}, k \in \mathcal{K}}$, and thus can be efficiently solved.

Fig. 3.10 Decomposition structure of problem (JTCOO-M)

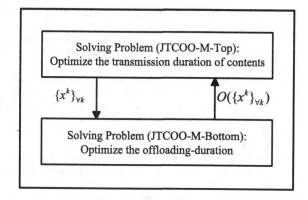

3.7 Summary

In this chapter, we have proposed an optimal resource allocation for the MUs' D2D-assisted traffic offloading for downlink content distribution. Specifically, we formulated a joint optimization of the content transmission and the MUs' offloading strategies, with the objective of minimizing the overall radio resource usage for downlink content distribution. Due to the MUs' limited transmit powers and energy budgets, the content transmission strongly influences the MUs' offloading durations, which makes the joint optimization problem very difficult to solve. To tackle with this difficulty, we exploited the decomposable structure of the joint optimization problem and characterized all possible cases for achieving the optimum. We then analytically derived the optimal solution for each of the cases and proposed an efficient algorithm to find the globally optimal offloading solution. Numerical results have validated the accuracy and computational efficiency of our proposed algorithm to compute the optimal offloading solution. Moreover, the numerical results also show that the optimal MUs' D2D-assisted offloading solution can significantly reduce the total radio resource usage for content distribution compared with other heuristic offloading schemes.

References

1. CISCO, "Cisco visual networking index: global mobile data traffic forecast update, 2015–2020 white paper," http://www.cisco.com/c/en/us/solutions/collateral/service-provider/visual-networking-index-vni/mobile-white-paper-c11-520862.html.
2. D. Camps-Mur, "Device-to-Device communications with WiFi direct: Overview and experimentation," *IEEE Wireless Communications*, vol. 20, no. 3, pp. 96–104, 2013.
3. A. Pyattaev, K. Johnsson, S. Andreev, and Y. Koucheryavy, "3GPP LTE traffic offloading onto WiFi direct," in *Proc. of IEEE WCNCW*, Shanghai, China, Apr. 2013.
4. QUALCOMM, "LTE advanced evolving and expanding into new frontiers," https://www.qualcomm.com/documents/lte-advanced-evolving-and-expanding-new-frontiers.

5. C. Sankaran, "Data offloading techniques in 3GPP Rel-10 networks: a tutorial," *IEEE Communication Magazine*, vol. 50, no. 6, pp. 46–53, 2012.
6. P. Phunchongharn, E. Hossain, and D. Kim, "Resource allocation for Device-to-Device communications underlaying LTE-advanced networks," *IEEE Wireless Communications*, vol. 20, no. 4, pp. 91–100, 2013.
7. Y. Wu, J. Chen, L. Qian, J. Huang, and X. Shen, "Energy-aware cooperative traffic offloading via Device-to-Device cooperations: An analytical approach," *IEEE Transactions on Mobile Computing*, vol. 16, no. 1, pp. 97–114, Jan. 2017, DOI:10.1109/TMC.2016.2539950.
8. J. Chen, Y. Wu, L. Qian, H. Peng, and H. Zhou, "Energy-efficient content distribution via mobile users cooperations in cellular networks," *Peer-to-Peer Networking and Applications*, 2016, DOI:10.1007/s12083-016-0519-3.
9. L. Al-Kanj, Z. Dawy, W. Saad, and E. Kutanoglu, "Energy-aware cooperative content distribution over wireless networks: Optimized and distributed approaches," *IEEE Transactions on Vehicular Technology*, vol. 62, no. 8, pp. 3828–3847, 2013.
10. L. Al-Kanj, V. Poor, and Z. Dawy, "Optimal cellular offloading via Device-to-Device communication networks with fairness constraints," *IEEE Transactions on Wireless Communications*, vol. 13, no. 8, pp. 4628–4643, 2014.
11. T. Wang, L. Song, Z. Han, and B. Jiao, "Dynamic popular content distribution in vehicular networks using coalition formulation games," *IEEE Journal on Selected Areas Communications*, vol. 31, no. 9, pp. 538–547, 2013.
12. L. Gao, G. Iosifidis, J. Huang, and L. Tassiulas, "Hybrid data pricing for network-assisted user-provided connectivity," in *Proc. of IEEE INFOCOM*, Toronto, Canada, Apr. 2014.
13. L. Vu, K. Nahrstedt, I. Rimac, V. Hilt, and M. Hofmann, "iShare: Exploiting opportunistic ad hoc connections for improving data download of cellular users," in *Proc. of GLOBECOM'2010*, Miami, America, Dec. 2010.
14. D. Niyato, P. Wang, W. Saad, and A. Hjorungnes, "Controlled coalitional games for cooperative mobile social networks," *IEEE Transactions on Vehicular Technology*, vol. 60, no. 4, pp. 1812–1824, 2011.
15. S. Dimatteo, P. Hui, B. Han, and V. O. Li, "Cellular traffic offloading through WiFi networks," in *Proc. of IEEE MASS'2011*, Valencia, Spain, Oct. 2011.
16. Y. Li, G. Su, P. Hui, D. Jin, L. Su, and L. Zeng, "Multiple mobile data offloading through delay tolerant networks," in *Proc. of ACM Mobicom'2011, CHANTS Workshop*, Las Vegas, NV, Sep. 2011.
17. Y. Li, Y. Jiang, D. Jin, L. Su, L. Zeng, and D. Wu, "Energy-efficient optimal opportunistic forwarding for delay-tolerant networks," *IEEE Transactions Vehicular Technology*, vol. 59, no. 9, pp. 4500–4512, 2010.
18. X. Wang, M. Chen, Z. Han, T. Kwon, and Y. Choi, "Content dissemination by pushing and sharing in mobile cellular networks: An analytical study," in *Proc. of IEEE MASS'2012*, Las Vegas, NV, Oct. 2012.
19. J. Whitbeck, M. Amorim, Y. Lopez, J. Leguay, and V. Conan, "Relieving the wireless infrastructure: When opportunitic networks meet guaranteed delays," in *Proc. of IEEE WoWMoM*, Lucca, Italy, Jun. 2011.
20. N. Golrezaei, A. Molisch, A. Dimakis, and G. Caire, "Femtocaching and Device-to-Device collaboration: A new architecture for wireless video distribution," *IEEE Communs. Magazine*, vol. 51, no. 4, pp. 142–149, 2013.
21. Y. Bi, H. Shan, X. Shen, N. Wang, and H. Zhao, "A multi-hop broadcast protocol for emergency message dissemination in urban vehicular ad hoc networks," *IEEE Transations on Intelligent Transportation Systems*, vol. 17, no. 3, pp. 736–750, 2016.
22. D. Ng, E. Lo, and R. Schober, "Energy-efficient resource allocation in multiuser OFDM systems with wireless information and power transfer," in *Proc. of WCNC*, Shanghai, China, Apr. 2013.
23. J. Wu, Y. Bao, G. Miao, and Z. Niu, "Base station sleeping and power control for bursty traffic in cellular networks," in *Proc. of ICC*, Sydney, Australia, Jun. 2014.
24. C. Li, J. Zhang, and K. Letaief, "Energy efficiency analysis of small cell networks," in *Proc. of ICC*, Budapest, Hungary, Jun. 2013.

25. B. Debaillie and C. Desset, "Power modeling of base stations," http://wireless.kth.se/5green/wp-content/uploads/sites/19/2014/08/BDebaille.pdf.
26. G. Auer, V. Giannini, C. Desset, I. Godor, P. Skillermark, M. Olsson, M. Imran, D. Sabella, M. J. Gonzalez, O. Blume, and A. Fehske, "How much energy is needed to run a wireless network?" *IEEE Wireless Communications*, vol. 18, no. 5, pp. 40–49, 2011.
27. A. Carroll and G. Heiser, "An analysis of power consumption in a smartphone," in *Proc. of USENIX Annual Technical Conference*, Boston, America, Jun. 2010.
28. A. Gupta and P. Mohapatra, "Energy consumption and conservation in WiFi based phones: A measurement-based study," in *Proc. of IEEE SECON*, San Diego, CA, Jun. 2007.
29. E. Weisstein, "Lambert W-function, mathworld," http://mathworld.wolfram.com/LambertW-Function.html.

Chapter 4
Conclusions and Future Directions

In this chapter, we summarize the brief and provide future research directions.

4.1 Conclusions

In this brief, we illustrate the importance and design methodologies of the optimal radio resource allocations for traffic offloading in heterogeneous small-cell networks. Specifically, we present two design examples corresponding to the two most important paradigms of traffic offloading, namely, the small-cell-based traffic offloading and the D2D-assisted traffic offloading. Based on the discussions and analysis throughout this brief, we present the following conclusive remarks.

- Traffic offloading through small-cell networks is an effective and cost-efficient manner to accommodate the rapidly growing traffic demand in cellular systems. Traffic offloading not only effectively relieves the traffic burden and congestion in macrocells, but also brings multi-folded benefits, such as enhancing throughput and coverage, improving users' experienced quality of services, and improving radio resource utilization. Nevertheless, to achieve these promising benefits, we need proper radio resource managements to support the traffic offloading in heterogeneous small-cell networks.
- Small-cell-based traffic offloading with dual connectivity is an efficient traffic offloading paradigm in heterogeneous small-cell networks. The key feature of traffic offloading via dual connectivity is that it allows mobile users to simultaneously schedule traffic to the macrocell and offload traffic to the small cells, thus yielding flexible traffic scheduling and efficient radio resource allocations. However, aggressive traffic offloading will lead to excessive co-channel interferences among mobile users, which compromise the benefits of traffic offloading. Hence, to optimize the benefit of traffic offloading while mitigating the adverse

© The Author(s) 2017
Y. Wu et al., *Radio Resource Management for Mobile Traffic Offloading
in Heterogeneous Cellular Networks*, SpringerBriefs in Electrical
and Computer Engineering, DOI 10.1007/978-3-319-51037-8_4

co-channel interference, we need to execute a joint optimization of uplink traffic scheduling and power allocations, e.g., in order to minimize the users' total mobile data cost. Efficient algorithms, by exploiting the hidden monotonic structure of the joint optimization problem, have been proposed to compute the optimal offloading solution. We have performed extensive numerical results to show that the optimized traffic offloading solution can effectively reduce mobile users' total data cost and enhance the overall offloading capacity.

- D2D-assisted traffic offloading is an efficient offloading paradigm (especially for downlink content distribution) by exploiting mobile users' D2D cooperation. Thanks to close-proximity of mobile users, the D2D-assisted offloading can effectively reduce cellular-link usage and improve radio resource utilization. However, due to users' limited transmit-powers and energy capacities, the content transmission and users' offloading strategies are strongly coupled, which together influence the overall offloading performance. Hence, to optimize the overall offloading performance (e.g., to minimize the overall radio resource usage), we need to execute an optimization problem that jointly optimizes the content transmission and the users' offloading strategies. Efficient algorithms, by exploiting the layered structure of the joint optimization problem, have been proposed to compute the optimal D2D-assisted offloading solution. We have performed extensive numerical results to show that the optimal D2D-assisted offloading can effectively reduce the overall radio resource usage compared with other offloading schemes.

4.2 Future Research Directions

This brief presents the results on optimal resource allocation for traffic offloading in heterogeneous cellular networks. The studies in Chaps. 2 and 3 can be further extended in several interesting directions, and we illustrate two promising directions as follows.

Resource allocation for traffic offloading powered by renewable energy: An interesting future direction is to investigate traffic offloading powered by renewable energy. Nowadays cellular systems and tremendous number of mobile devices have consumed a huge amount of carbon-based energy and contributed to a considerable amount of carbon emission. Thanks to the low-power nature of small cells, exploiting renewable energy to feed small cells (especially microcells and picocells) has emerged as an effective and efficient approach to reduce conventional carbon-based energy consumption and achieve sustainable and environmentally friendly development [1, 2]. Many studies have showed the viability and benefits of exploiting renewable energy to power small cells [3–5]. Recent technological advances for energy harvesting devices and energy storage systems have further facilitated the deployment of exploiting renewable energy by small cells. Hence, traffic offloading through renewable energy powered small cells becomes a promising paradigm that jointly exploits the transmission capacity provided by small cells and renewable energy provided by environments. One of the most important issues in exploiting renewable energy is the intermittency of renewable energy supply. For instance, energy collected by wind turbines might vary temporally and geographically due to varying weather conditions. Hence, if small cells are powered by renewable energy, the varying availability of renewable energy will influence the cells' offloading capabilities (for instance, when the stored energy level is low, small cells need to slow down offloading rate or reduce the number of admitted mobile users). Accordingly, the traffic offloading strategies (e.g., admission control, traffic scheduling, resource allocation, and mobility management) need be dynamically adjusted according to energy source dynamics, in addition to many other factors such as the users' traffic demands and QoS requirements. To tackle with the intermittency in renewable energy supply, the hybrid use of conventional on-grid power supply and renewable energy supply to power small cells has been widely considered (e.g., in [6]). Specifically, when the stored energy level from harvesting renewable energy is low, small cells can switch to on-grid power supply to offload users' traffic. Moreover, efficient schemes to accurately predict the availability of renewable energy supply become very important.

Mobility-aware resource allocation for traffic offloading: The studies in Chaps. 2 and 3 mainly focus on the relatively static scenario (e.g., the users keep static within an area or move slowly). This scenario corresponds to the most prevalent case for traffic offloading in cellular networks. However, an important future direction is to take into account users' (fast) mobility and investigate the corresponding mobility-aware radio resource allocation for traffic offloading. Since the fast-moving users may frequently encounter the boundaries of small cells, we need to jointly optimize the radio resources of multiple neighboring cells during the users' offloading process, such that the moving users can flexibly and adaptively offload traffic when crossing different cells without suffering from a degraded quality of service. In particular, information such as the moving users' mobility pattern and the neighboring cells' resource utilization will be crucial to the design of mobility-aware resource allocation for traffic offloading.

References

1. R. Martin, "Nearly 400,000 off-grid mobile telecommunications base stations employing renewable or alternative energy sources will be deployed from 2012 to 2020," http://www.navigantresearch.com/newsroom/nearly-400000-off-grid-mobile-telecommunications-base-stations-employing-renewable-or-alternative-energy-sources-will-be-deployed-from-2012-to-2020.
2. M. Ismail, W. Zhuang, E. Serpedin, and K. Qaraqe, "A survey on green mobile networking: From the perspectives of network operators and mobile users," *IEEE Communications Surveys and Tutorials*, vol. 17, no. 3, pp. 1535–1556, 2015.
3. H. Dhillon, Y. Li, P. Nuggehalli, Z. Pi, and J. Andrews, "Fundamentals of heterogeneous cellular networks with energy harvesting," *IEEE Transactions on Wireless Communications*, vol. 13, no. 5, pp. 2782–2797, 2014.
4. G. Piro, M. Miozzo, G. Forte, N. Baldo, L. Grieco, G. Boggia, and P. Dini, "HetNets powered by renewable energy sources: Sustainable next-generation cellular networks," *IEEE Internet Computing*, vol. 17, no. 1, pp. 32–39, 2013.

5. H. Yang, J. Lee, and T. Quek, "Heterogeneous cellular network with energy harvesting-based D2D communication," *IEEE Transactions on Wireless Communications*, vol. 15, no. 2, pp. 1406–1419, 2016.
6. S. Zhang, N. Zhang, S. Zhou, J. Gong, Z. Niu, and X. Shen, "Energy-aware traffic offloading for green heterogeneous networks," *IEEE Journal on Selected Areas in Communications*, vol. 34, no. 5, pp. 1116–1129, 2016.

Printed in the United States
By Bookmasters